The Physics of Stars

The Physics of Stars

S. A. Kaplan

Formerly of
Department of Physical and Mathematical Sciences,
Gorky University, USSR

Translated by
Renata Feldman
Nice Observatory, France

JOHN WILEY & SONS

Chichester · New York · Brisbane · Toronto · Singapore

First published as *Fizika zvezd*, third edition, in the USSR in 1978.

Copyright © 1982 by John Wiley & Sons, Ltd.

All rights reserved.

Library of Congress Cataloging in Publication Data:

Kaplan, S. A (Samuil Aronovich)
  The physics of stars.
  Translation of: Fizika zvezd. 3rd ed.
  Includes index.
  1. Stars.   I. Feldman, Renata.   II. Title.
QB801.K2813   523.8   82-2651

ISBN 0 471 10327 6   AACR2

British Library Cataloguing in Publication Data:

Kaplan, S. A.
  The physics of stars.
  1. Stars
  I. Title   II. Fizika zvezd. *English*
  523.8   QB801

ISBN 0 471 10327 6

Typeset by Activity, Salisbury, Wilts
and printed by the Pitman Press Ltd., Bath, Avon.

Contents

Preface

This translation has been made from the Russian third edition of Kaplan's *The Physics of Stars*, which was essentially rewritten. Some of the general questions in the first chapters have been given in less detail which allowed us to reduce the volume of the book. On the other hand, we have introduced a new chapter treating neutron stars and 'black holes' and the generation and formation of stars has been completed. We have taken into account the latest achievements in astrophysics.

From the foreword to the first edition

Astronomy, certainly the oldest science, is today undergoing a real revolution. Everybody knows that the achievements in space have opened new aspects to man, allowing us, for the first time in history, to realize interplanetary experiments. Radioastronomy has yielded powerful new methods for the study of the Sun, the galaxies, and the interstellar medium. In addition there have been notable achievements in optical astronomy.

However, the most important fact is that astronomers can now use for the study of stellar objects the voluminous experimental and theoretical material gathered from modern physics, especially from atomic and nuclear physics. We can now understand and study problems which have been considered unsolvable for centuries: to know the temperature in the interior and on the surface of stars, to determine the chemical composition of celestial bodies, to find the source of stellar energy, to study the stellar evolution, etc., etc. The key to all these problems is in physics.

In this book you will learn how to study the structure of stars with the help of school physics and how to understand the processes taking place in stellar interiors—to study stellar evolution. You will also learn how to calculate the characteristics parameters of stars, for example their temperatures, and, if their masses and radii are given, to find the oscillation periods of variable stars with their mean density, to calculate the molecular weight of the stellar matter, to find the magnitude of stellar energy sources, etc. The results will not have the precision needed for modern science but will be sufficient to give a good understanding of stars.

You will find here neither a description of the beauties of the constellations nor a comparison of stellar dimensions with Earthly objects. To us, stars, although beautiful, are objects of physical study. Different subjects will be studied in the same way as in laboratories: electric fluxes, magnetic fields, and other physical phenomena will be studied with a pen and a piece of paper.

It should be noted that formulae are given here. They may be few but they play an important part, forming the basis of the text. Using these formulae we will learn how to understand and estimate the conditions in stellar interiors. Some of them have already been met in school physics; others which may be new are thoroughly explained.

Although this book is called *The Physics of Stars* we do not deal with stellar physics in general. We have chosen one chapter of this science which is quite large. It is the theory of stellar structure and stellar evolution. Why have we chosen these questions and stars as objects for our study of celestial bodies with the help of physics? The reasons are diverse. The first reason is the importance of this question: the stars represent the foundation of the Universe (the greatest part of cosmic material is contained inside of stars) and the understanding of stellar evolution is very important for the materialistic vision of the World. Second, the physical theory of inner stellar structure is relatively simple. It may seem a paradox but we know more about the inner structure of any star, even far away, than about the inner structure of our Earth or the surface of the Sun, which we have been studying for many decades.

Although we can never actually reach the interior of the Sun or of the stars we can quite easily calculate their structure and find out the physical conditions of the stellar matter. These conditions appeared to be so simple, although very different from those on Earth, that one of the founders of the inner stellar structure theory, the English astrophysicist A. Eddington, said: 'There is nothing more simple than a star.'

For this reason a knowledge of school physics is sufficient for the study of the inner structure of stars. This would no longer be the case if we were to treat stellar atmospheres.

1

Fundamental characteristics of stars

When approaching the physical study of a body we must first of all characterize it, i.e. determine its mass, form, dimension, aggregate state, chemical composition, etc. Likewise, when approaching the physical study of stars we must first of all determine the basic values characterizing stars, in other words, the parameters of stars.

Many parameters exist to describe the different characteristics of stars, but for the theory of the inner structure of stars and for the study of stellar evolution only a few fundamental parameters are important: the mass, radius, luminosity, spectral class, chemical composition. These parameters are determined by observation. For some of them this is easy to do, but for other more complicated and difficult methods are needed. Finally, it is apparent that it is possible to determine an important number of fundamental parameters for only a comparatively few stars (and all of them for the Sun only). This certainly is a serious difficulty for the theory, but the known parameters are sufficient to make a great number of important deductions.

The aggregate composition of stars and their form do not require elucidation. Stars are big gas spheres (sometimes they have the form of an ellipsoid). Probably the most important parameter of stars is their mass, but first we shall look at the other paremeters.

Primarily, what do we observe when looking at stars? It is their brightness, measured in so-called stellar magnitudes. One of the fundamental characteristics of stars is the capability of emitting energy—to shine. Therefore the first stellar parameter we shall introduce is 'luminosity'. The luminosity of a star, designated by the letter L and measured in ergs per second, is the quantity of light emitted by the total surface of the star in one second.

The luminosity of the Sun can easily be determined. We know that each square centimetre of the Earth's surface receives from the Sun, in the form of light, about two calories every minute. This quantity, called the solar constant, has been determined many times with the help of special instruments called actinometers. As the Sun emits energy uniformly in all directions, we can multiply the two calories by the area of the surface of a big sphere with its centre at the Sun and a radius equal to the distance between the Earth and the Sun. Thus we obtain the amount of radiation yielded by the Sun in the form of light in one minute. Converting the calories into ergs and the minutes into seconds we find the brightness of the Sun in ergs per second. We let the reader perform his own calculations and give here only the final result: the luminosity of the Sun $L_{\odot}$ is equal to 3.8×10^{33} erg/s.

In order to determine the luminosities of other stars we must compare the light they give on Earth with the light coming from the Sun. It is true that the comparison of these two quantities is in a sense a comparison of day and night. However, this difference is due only to the fact that the Sun is close while the stars are far away (the closest star is 250 000 times farther away than the Sun) and we know that the emitted radiation is inversely proportional to the square of the distance. We must therefore take into account the fact that the stars are at different distances from the Earth. Thus, in order to compare the different luminosities of stars we must mentally move all the stars, including the Sun, to the same distance from the Earth, equal to 3.08×10^{14} km or 10 pc (parsec). (A parsec is the distance at which the radius of the Earth's orbit is seen at an angle of one second of angle. Consequently, at a distance of 10 pc the semi-major axis of the Earth's orbit is seen at an angle of under 0.1″. At this distance the Sun would appear to us as a very faint star not easily seen in the sky.) Once the light from the stars and the Sun is determined by observation, and knowing that the radiation varies inversely with the square of the distance, we can calculate what would be the light from the Sun and the stars at a distance of 10 pc. Since the ratio of the light from the sources situated at the same distance equals the ratio of their luminous intensities, i.e. of their luminosities, we can determine the ratio of the luminosities between a star and the Sun and consequently the luminosity of the star in ergs per second.

We shall now consider other stellar parameters. By carefully observing stars we can easily see that they have different colours. The colour of the Sun is yellowish while that of the brightest star in our sky—Sirius—is white. The majority of stars have a more reddish colour than the Sun. One would imagine that the difference in the colour of the stars is explained by their different surface temperatures.

It is well known that if we heat, for example, iron, as the temperature increases it becomes first dark red, then yellow, and finally white and incandescent. An expert blacksmith is easily able to determine the temperature by the colour of the heated metal. This is also true of stars: the difference in colour of stars is due to the difference in temperature and the astronomer must know how to determine the temperature by the colour of the star. The relationship betwen the colour and the temperature of a heated body is known in physics under the name of Wien's displacement law: $\lambda_m = 0.29/T$. T is the temperature of the heated body on the absolute temperature scale and λ_m is the wavelength (in centimetres) of the light emitted by this body at its maximum intensity. For example, this formula gives at a temperature $T = 4000$ K a maximum of radiation of $\lambda_m = 7.2 \times 10^{-5}$ cm, i.e. the wavelength of the colour red. Consequently, the temperature at the surface of a red star is about 3000–4000 K. The temperature of yellow stars is 5000–6000 K. (In particular, the temperature of the solar surface is 5760 K). The hotter stars ($T = 10\ 000$ K and more) are of a whitish colour with a bluish shadow. For these hot stars the maximum of radiation is in the ultraviolet region of the spectrum ($\lambda_m < 3 \times 10^{-5}$ cm) which is not perceptible to the eye. Therefore we only observe the less intensive radiation of the stellar surface with wavelengths in the visible region of the spectrum (blue, yellow, red). The sum of this radiation gives the whitish colour of the stellar surface.

Astronomers often determine the temperature of a stellar surface from its colour, but if it is possible to obtain the stellar spectrum the temperature can be determined with greater precision. We remind the reader that from laboratory observations of emission or absorption spectra and from compiled lists it is possible to determine from the position of the spectral lines the presence of any chemical element in the given matter and from the intensity of the line the number of atoms of these elements.

We shall first give some of the results obtained from the study of the chemical composition of stars.

It became apparent that the upper atmospheric layers of the great majority of stars have approximately the same chemical composition (this does not mean that in the interior of the stars the chemical composition is the same), but there are also anomalies. The most common elements observed in stellar atmospheres are helium and hydrogen. For example, in the atmosphere of the Sun for each atom of oxygen there are nearly two thousand hydrogen atoms and about two hundred helium atoms. There is about three times less nitrogen than oxygen and five times less carbon (when considering the number of atoms). There is a million times less lithium and beryllium than oxygen. The amount of neon is approximately the same as that of oxygen and that of magnesium and argon is about ten times smaller. Even smaller are the amounts of iron (about thirty times less), silicon (about fifty times less), and chlorine (about fifteen times less). The abundance of other elements is very low—a hundred, thousand, and million times less than oxygen. An entirely defined law exists, although with many exceptions: the heavier the chemical element the lesser its amount. This law is also observed on Earth and in many respects the chemical composition of stars and that of the Earth is similar. However, there are also very distinct differences: on Earth we have much oxygen and silicon and little hydrogen while stars have large amounts of hydrogen and helium; also on Earth there is very little of the inert gases which exist in greater quantities in the stars. These facts are very important for the theory of stellar evolution.

We must also note that for a series of stars, or even groups of stars, there exist great anomalies in chemical composition. For example, there exist so-called 'metallic stars' in which there are more metals than in ordinary stars, there are stars with a great abundance of rare-earth elements, there are carbon stars with a high abundance of carbon, and there are stars containing an element, technetium, which is generally unstable and not found on Earth or on the Sun in a natural state. In the case of the so-called 'subdwarfs' there is an even greater relative amount of hydrogen and helium. These differences in chemical composition are very important for the study of stellar evolution and indicate, to a certain extent, the accuracy of our assumptions on the structure of the star. We shall give one example here. The light nuclei of lithium, beryllium, and boron are involved, relatively easily, in thermonuclear reactions, in the process of which these nuclei disappear (in the end they are converted into helium) and are not renewed again, which explains their small abundance in stars. All these elements could exist in the original matter from which the stars were formed. Therefore one could expect the light elements to be burned in the stellar interiors while being maintained in the outer layers. This is very important—the stellar matter does not intermix. In cold stars the stellar matter does intermix, which we shall find out when studying the structure of these stars. Therefore there is almost no lithium, beryllium, or boron. In hot stars there may be a greater amount of lithium. Unfortunately the solution of this 'lithium problem' is very difficult, since it is not easy to determine the abundance of rare elements in stellar atmospheres.

Apparently, in the case of 75 per cent. of stars of the spectral classes F5–G1 there is more than five times the amount of lithium in the Sun, whereas the number of stars of the classes G2–G8 with this abundance of lithium is only 25 per cent. Probably the mixing is more important here than in the case of hotter stars, but smaller than in the case of cool stars.

Thus, the chemical composition is also a parameter of the star. As we shall see later, in most cases it is not important to know the detailed values of the relative abundance of all elements. For the construction of the theory of stellar evolution one must know the amount of hydrogen, helium, and of heavy elements taken all together. The relative content of hydrogen is denoted by the letter X, that of helium by Y, and that of heavy elements (chiefly carbon, nitrogen, oxygen) by Z. It is evident that $X + Y + Z = 1$.

Let us consider now how to determine the temperature of a stellar surface by spectral analysis. We shall illustrate this with the following example. We already know that hydrogen is the most abundant element in stars. Therefore we would expect that the hydrogen lines would be the most intense lines in stellar spectra. However, in fact this is not the case. The spectral lines of hydrogen are indeed very intense in stars with a surface temperature of about 10 000 K, but in red stars (with a surface temperature of 3000–4000 K) or in very hot blue stars with a surface temperature above 15 000–20 000 K the hydrogen lines are almost invisible. This phenomenon can be easily explained. It is well known that atoms are composed of nuclei and electrons moving around the nuclei in defined orbits and that when a quantum of light or photon is absorbed the electrons jump from a lower orbit to an upper one. In order to effect such a transition from one orbit to another the photon must have a specific energy. It is easy to understand that in the atmospheres of cool stars, with low surface temperatures, the energy of the particles and the photons is small and is not sufficient to compel the electrons in hydrogen atoms to jump from one orbit to another. Therefore, in the atmospheres of these stars the majority of hydrogen atoms neither absorbs nor emits photons. The hydrogen lines in the spectra of these stars are weak. On the other hand, on the surface of very hot stars the energy of the particles and photons is so great that they tear off the electrons in hydrogen atoms, or as we say they ionize them. Certainly the nuclei of hydrogen atoms, having lost electrons, cannot absorb or emit photons. It is true that sometimes they capture free electrons—we call this process 'recombination'—and are then capable of emitting and absorbing light. This rare phenomenon, and the collision with the 'energetic' photon or particle which follows, again tears off the electrons and ionizes the atoms. Consequently, in the spectra of hot stars the hydrogen lines are also weak.

Only at a stellar surface temperature of about 8000–10 000 K is the number of non-ionized hydrogen atoms and the energy of the photons sufficient to allow frequent electron transitions in these atoms and, therefore, to form intensive hydrogen lines in the spectra. Indeed, the same arguments are valid for atoms of other chemical elements. Since each element has its own particular spectral lines, the presence in the spectrum of such spectral lines reflects not only its chemical composition but also the surface temperature. Moreover, since the chemical composition of stars is approximately the same (with rare exceptions) the difference in the stellar spectra is, in the first instance, exactly determined by the different temperatures on their surfaces.

Long ago, when this property was not yet known, astronomers classified stellar spectra by denoting the different types of spectra with Latin letters and by setting them in a defined, so-called spectral sequence. We now know that the order of the stellar spectra in this sequence is determined by the temperature of their surface.

Initially the spectral sequence was determined according to the order of the Latin alphabet. However, with the discovery of the dependence of this sequence on the temperature, this order has had to be changed and now students have contrived mnemonic laws for memorizing the spectral sequence. The best way is probably the English sentence: Oh Be A Fine Girl Kiss Me Right Now. Some of the letters in the initial sequence disappeared when it turned out that the corresponding spectra were not related to stars but either to nebulae or, if they referred to stars, to already known spectral classes notable only for anomalies in the spectrum. The final form of the spectral sequence is:

$$R—N$$
$$O—B—A—F—G—K—M$$
$$S$$

blue   white   yellow   red

In the spectra of class A stars (initially this was the beginning of the sequence) the hydrogen lines are the most intensive. In classes F, G, K, and M the hydrogen lines become successively weaker. In classes B and O the hydrogen lines also become weaker but, as we have seen above, this is for another reason. Therefore, we have changed the initial order. In the stellar spectra of classes R, N, and S there are almost no hydrogen lines. There are very few stars in classes R, N, and S and apparently they have a slightly different chemical composition—for this reason they branch off from the common spectral sequence. In stars of classes R and N there is probably more carbon (they are often called carbon stars). In the stellar spectra of class S one observes zirkonium oxide lines, whereas in the spectra of most prevailing stars of classes K and M the lines of titane oxide dominate. One must not think that titanium is to be found in these stars in large amounts. The fact is simply that at this temperature the titanium oxide is more easily found in the lines of the visible part of the spectrum than other elements or chemical combinations. One must also pay attention to the fact that at low temperatures (spectral classes K, M, R, etc.) there are still chemically combined molecules, while on the surface of hotter stars (spectral classes O to G) all molecules are dissociated into atoms. According to the presence or absence of molecules or lines in the spectra, one can also evaluate the temperatures of stellar surfaces.

We must note that the spectral classification described above (distribution of stars according to the spectra within ten classes) has proved to be too rough. For this reason astronomers have divided each interval of this sequence into ten parts. For example, there are stars with spectral classes B0, B1, B2, ..., etc., up to B9. Then follow A0, A1, ..., A9, F0, ..., etc. Stars with a large number have a small surface temperature. Moreover, a more thorough study of the spectra has revealed more subtle differences. The stellar spectra depend not only on temperature and chemical composition but also on the dimension of the star, more precisely, on the density of the gas in its atmosphere (stars of smaller dimension have denser atmospheres). For this reason, to designate the spectral class of the stars is also, if necessary, an index d denoting the spectrum of a relatively small star (from the English word 'dwarf'), or g (giant) denoting the spectrum of big stars, or c denoting the spectra of very big stars called supergiants; for example dM3, gF5, cB8, etc.

We have now discussed two parameters of the stars—the luminosity and the spectrum (or temperature). Right away there is a question: can a star of a given spectral class have an arbitrary luminosity (i.e. emit an arbitrary amount of energy) and, inversely, can a star with a given luminosity have an arbitrary surface temperature? This question is very important for the comprehension of stellar evolution.

In order to answer this question, we shall construct a so-called spectrum–luminosity diagram (such diagrams were composed for the first time by the Dutch astronomer Herzsprung and the American astrophysicist Russel in the years 1905–1913 and therefore are often called the Herzsprung–Russel diagrams). In order not to deal at once with a great

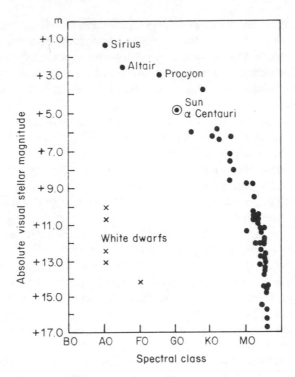

Figure 1 Spectral–luminosity diagram for stars situated within 5 pc from the Sun

number of stars, we shall choose the stars closest to us (situated not farther than 5 pc) and shall mark them by points on a graph on which the ordinate indicates the logarithms of luminosity and the abscissa the spectral classes.

As a result we obtain the diagram shown in Figure 1, from which it is seen at once that the luminosity of the star and its spectral class are related by a determined, though not unique, dependence. The majority of stars is situated along a line going from the hot and brilliant stars to the cold and faint stars. This is the well-known main sequence, to which the greater majority of stars including our Sun (spectral class G2) belongs. Note that there are many more cold and faint stars than hot and brilliant ones, but because the latter stand out better in the sky they have been given genuine names.

In the lower left-hand part of the diagram the faint but hot stars are situated, which, according to now understandable reasons, are called white dwarfs (in contrast, the faint stars of the main sequence are called red dwarfs). The white dwarfs represent a large group of stars. They are distinctly different from other stars by their structure and we shall study them separately.

We can see that the luminosity of the star and its spectral class are related. One of the first tasks of the theory is to explain this dependence, to find the basic physical phenomena. How this has been done by modern astrophysics will be explained later. At this point we shall note that it was only after construction of the diagram that a sense of evolution was attached to it. It had been assumed that the stars evolved along the main sequence from the hot and brilliant stars towards the cold and faint ones. It then turned out that the stellar evolution had a more complex character than the representation of stars showing 'early' stars in the upper left-hand part of the diagram and 'late' stars at the other end of the main sequence.

The spectral–luminosity diagram shown in Figure 1 does not appear to be entirely 'representative'. In its construction we limited ourselves to the close stars only. In particular, very bright but far away stars are not represented. One can certainly construct such a diagram for all the stars for which the spectrum (or the colour and the luminosity) can be determined. This is the way it has been done before. Such diagrams are representative in the sense that one can see in them other groups of stars that do not occur in the diagram of Figure 1 (red giants, supergiants, subgiants and subdwarfs, etc). However they are less representative in another respect for stars of different ages and different origins are mixed 'into one lot'. We could construct spectrum–luminosity diagrams for separate stellar systems with stars of the same age. In spite of the fact that on such diagrams there are less stars, they are more 'representative'. We shall consider such diagrams below.

A fundamental parameter of a star—its radius—can easily be determined from the given luminosity and surface temperature if the Stefan–Bolzmann radiation law is used. According to this law each square centimetre of the surface of a body heated to T K emits, in all directions, a light energy (visible, ultraviolet, infrared rays) of σT^4 ergs per second. Here σ is the Stefan–Bolzmann constant and its numerical value is 5.7×10^{-5} erg/(cm^2 s deg^4). Note that with the increase of temperature the intensity of the radiation of the body increases proportionally to the fourth degree of temperature. You can verify this on your own if you remember that iron radiates strongly when heated to a temperature of 600 K (twice as much as room temperature in the Kelvin scale).

If we designate the radius of a star by R (in centimetres) then its total surface equals $4\pi R^2$ square centimetres. Each square centimetre emits σT_e^4 ergs per second, where T_e stands for the surface temperature of the star. Subsequently, the total emission of the star in one second, i.e. the luminosity, equals

$$(1) \quad L = 4\pi R^2 \sigma T_e^4$$

With this formula one can easily find the radius of the star if its luminosity (or the absolute bolometric stellar magnitude) and its surface temperature (or spectral class or colour index) are known. We have already seen how these parameters are determined. The radius of the Sun can be measured directly. It is seen from Earth with an angle of $16' = 0.0046$ rad. Multiplying this value by the distance between the Earth and the Sun (i.e. by 150 million km) we obtain for the radius of the Sun $R_\odot = 7 \times 10^{10}$ cm. The reader can easily verify that the above formula yields the same value for the radius of the Sun ($L_\odot = 3.8 \times 10^{33}$ erg/s, $T_\odot = 5760$ K).

We cannot measure directly the radii of other stars, since even with bigger telescopes we are unable to 'see' the disk of the star. It is true that for relatively close stars of large dimension it has been possible with the help of special interferometers (instruments with action based on interferential phenomena) to determine their angular diameters, but the precision of such measures is very small, usually less than in calculations with formula (1).

There is another possibility of determining stellar radii. We must suppose that a great number of stars, probably more than half, are not 'single'. By this we do not mean planetary systems such as that of our Sun, but so-called multiple stellar systems (composed of two, three, or more stars). For example, in the simple case of a binary stellar system both stars (they are called components) move under the action of a universal gravitation force in elliptic or circular orbits around a general centre of gravity—exactly as the planets rotate around the Sun. If the two components of the pair are always a great enough distance away from one another for us to see them separately, then this system is called visually binary. However, the stars are often too close to one another for us to see them separately, even with very powerful telescopes. We can determine that there are two stars and not one, either by the periodic changes in their spectra (spectrally binary) or by the eclipses of one star by the other (eclipsed variables). This last case will now be discussed.

Let us admit that the line of sight connecting the observer and the centre of the binary system is situated in the plane of the components' orbits of the pair. Then, during the rotation of the stars each one of them will from time to time eclipse the other one, in the same way as the Moon covers the Sun at the time of a solar eclipse. If we know the duration of the eclipse, the distance between the components of the pair, and the velocities of their motion, it is simple to calculate the dimension of the stars. It will be left to the reader to consider this problem on his or her own.

The study of eclipsed variables has revealed many interesting phenomena. For example, if both stars are situated so close to each other that their surfaces almost touch (as we say, they form a close pair), then the stars have an ellipsoidal form and resemble a melon, the elongated ends being turned one to the other. In the case of very close pairs the major axes of the stars are about 25 per cent. longer than the minor axes. This elongation is due to the attraction of the surface layers of one star by the other star. It is interesting to notice that the sides of the ellipsoidal stars facing one another are cooler than the rest of the surfaces.

We have probably spent more time than is necessary on the description of eclipsed stars, especially as the determination of the radii with formula (1) is still more accurate. However, in so doing we have come to know the very interesting phenomenon of ellipsoidal stars and have also studied binary systems (including the eclipsed system), practically the only way of determinating the stellar mass—the most important parameter which defines the structure and the evolution of stars.

For the determination of the masses of celestial bodies Kepler's third law is used: the squares of the rotation periods of planets are related as the cubes of their mean distances from the Sun. Newton completed this law by determining the coefficient of proportionality. In its complete form Kepler's law is

$$\frac{a^3}{P^2(M_\odot + m)} = \frac{f}{4\pi^3}$$

Here $M_\odot$ is the mass of the Sun in grams, m is the mass of the planet in grams, f is Newton's gravitation constant (in the CGS system $f = 1/15\ 000000\ \text{cm}^3/(\text{s g})$), a is the mean distance of the planet from the Sun in centimetres, and P is its rotation period in seconds. With this formula the mass of the Sun has been found. You can verify on your own the calculation by introducing the parameters for the motion of the Earth ($a = 150$ million km $= 15 \times 10^{13}$ cm, $P = 1$ year $= 3.1 \times 10^7$ s). The Earth's mass is small compared to the Sun's mass, so it can be neglected in formula (2). After a simple calculation we obtain $M_\odot = 2 \times 10^{33}$ g.

In binary stellar systems the stars move under the action of the same force of universal gravitation and consequently their motion obeys the same Keplerian laws, in particular the third one. We must now introduce in the place of $M_\odot + m$ in equation (2) $M_1 + M_2$—the sum of masses of both components. We must also know the distance between the components of the pair. Thus, knowing the distance between the components of the binary system and their period of rotation we can determine with this formula the total mass of the stars. With the total mass we can then estimate the mass of each component. At present the masses of many scores of stars are already known. Generally they are not very different from the mass of the Sun (from 0.1 to 50 $M_\odot$).

We have already examined the relationship between different stellar parameters. The relation between the luminosity and the spectrum (or surface temperature) is determined by the Herzsprung–Russel diagram. The relation between the luminosity and the radius yields no new information, as the stellar radius according to equation (1) is expressed by the luminosity and the surface temperature.

We can obtain a new relationship from the study of the luminosity and the mass insofar as the latter is determined in a quite independent way. We mark on the abscissa the logarithms of the stellar masses (or, even better, the logarithms of the ratio of stellar mass to mass of the Sun) and on the ordinate the logarithms of the luminosities. Thus we obtain the diagram shown in figure 2 which is called the mass–luminosity relationship. It shows only stars of the main sequence (the cross marks the Sun, the points the stars in spectroscopic binary systems, and the circles the stars in visual binary systems). We see that at least for a star in the main sequence the stellar luminosity and consequently also its spectrum are simply determined by the mass. This is a very important observational conclusion and the first problem

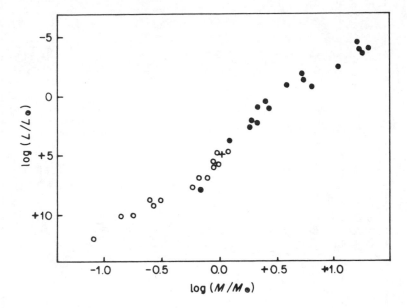

Figure 2 Mass–luminosity relationship. The cross marks the Sun, the points the spectroscopic binary stars, and the circles the visual binary stars

of the theory of inner stellar structure consists of finding the physical laws which define these relations.

In order to give the reader an image of the characteristic values of the stellar parameters we have shown them in Table 1 for a series of typical stars. There are also giants and supergiants, which we shall study later.

The stars shown in Table 1 are typical in the sense that they represent objects radiating their energy mainly in the visible spectral region. Such stars have been the object of study in 'classical' astronomy for tens and hundreds of years. Nevertheless, in recent times, thanks to the fantastic progress in new areas of astronomy (radio astronomy, X-ray astronomy), new types of objects have been discovered which, generally speaking, are also stars but of a quite different type—namely pulsars and 'black holes'. Certainly, in a study of the physics of stars we cannot leave aside these objects. Pulsars and 'black holes' also have luminosities, masses, and radii but their properties are so unusual that it is better to describe their parameters separately, at the same time as discussing the physics of these objects (see Chapter 6).

Let us come back to the 'ordinary' stars. We have already determined a few basic parameters: the mass M, the luminosity L, the radius R, the spectrum (or surface temperature T_e), the chemical composition (in particular the abundance of heavy elements Z). At once the question arises: do these parameters change with time for a given star? in fact, once formed, stars

Table 1 A few stellar parameters

Name of the star	Mass $M/M_\odot$	Luminosity $L/L_\odot$	Radius $R/R_\odot$	Surface temperature T_e, K	Spectral class
Main sequence					
ζ Arietis	10.2	220	3.5	11 000	B8
Vega	2.8	85	3.0	9 500	A0
Sirius	2.1	27	2.0	9 250	A1
Procyon	1.8	7.4	2.2	6 570	dF5
α Centauri	1.02	1.3	1.2	5 730	dG2
70 Ophiuchi	0.78	0.51	0.89	4 900	dK0
ξ Boötis	0.72	0.10	0.82	4 200	dK4
η Cassiopeiae	0.54	0.09	0.82	3 600	dM0
Kruger 60	0.26	0.007	0.26	3 000	dM3
Giants					
Capella	3.3	220	23	4 900	gG8
Arcturus	4.2	130	26	4 000	gK2
Aldebaran	4.0	360	45	3 800	gK5
Supergiants					
Rigel	40	2×10^5	138	11 200	cB8
Antares	19	3×10^4	560	3 300	cM1
White dwarfs					
40 Eridana	0.44	0.0035	0.0017	10 000	DA2
Sirius B	1.0	0.0027	0.02	8 200	DA5
Van Maanen's star	0.3	$\sim 2 \times 10^{-2}$	$\sim 10^{-2}$	8 000	DG

evolve, they consume their energy supply, which is by no means 'inexhaustible', and at a certain time the evolution of each star comes to an end. Obviously, if not all at least some of the parameters should change with time. For this reason we need one more important parameter—the age of the star.

It is possible to determine the age of stars through the study of stellar systems. We have already seen the simple form of stellar systems—the binary stars. The study of binaries, besides the definition of mass and radius, can also yield other valuable information. Apparently both components of the pair originated at the same time and therefore have the same age. Both components can be nearly identical stars but it is often the case that one of the stars belongs to the main sequence and the other to the white dwarfs group (as, for example, in the case of Sirius). Consequently, one can say that binary stars of the same age can be very different.

Besides the binary stellar system there exist more complex multiple systems composed of several stars (three, six, or even more), as well as so-called galactic star clusters and stellar associations, including scores and sometimes hundreds of thousands of stars. All these stellar systems enter the structure of the enormous stellar system—the Galaxy—with the amount of stars totalling some 150 milliards. Another example of a huge stellar system, of the same type as our Galaxy, is the well-known Andromeda Nebula. The galaxies, in their turn, amalgamate into clusters of galaxies. Finally, all these galaxies and their clusters taken all together form what we now call the supergalaxy.

It is very important to know that the Galaxy does not only represent one homogeneous stellar system but is composed of a great number of interpenetrating subsystems; we shall study them in detail below. Besides stars, the stellar systems are also composed of interstellar gas and the cosmic dust.

It is evident that stars belonging to the same stellar cluster have the same age. Probably stars which belong to the same galactic subsystem are also very close in age.

Can we find a concrete definition for the age of a star? Apparently it is possible. There are stellar systems which are very unstable and therefore should rapidly disintegrate. The characteristic time of collapse of such systems can be determined: it is of the order of the time each star needs to cross the diameter of the system; evidently the stars belonging to such systems are not old. The unstable systems of the type 'trapezium' are multiple stellar systems where all distances between the components are comparable, and associations are very scattered galactic clusters. In this way stars were discovered where the age is about a few hundred thousand years (which is very little in comparison with the age of the whole Galaxy—more then tens of milliards of years). In general, all galactic clusters observed now are comparatively young but their age is different. Globular star clusters are considerably older.

We now return to the colour–luminosity diagram. If we construct such a diagram for each cluster we can be sure that there will be only stars of the same age and the dependence of the luminosity on the spectrum will not be distorted by evolutionary factors. Figure 3 displays in one graph several spectra (or

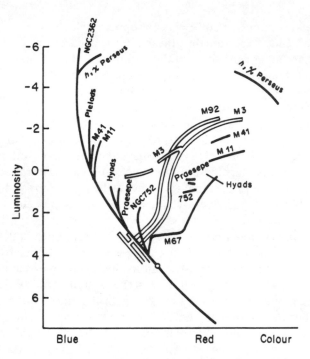

Figure 3. Summary colour–luminosity diagram for nine discovered scattered galactic clusters (thick lines) and for two globular clusters (thin lines). This diagram is very important for the theory of stellar evolution

colour–luminosity diagrams) for a series of galactic and globular clusters (obtained by Sandage). Here the thick lines represent the diagrams of galactic clusters and the thin lines the diagrams of globular clusters. Next to each line we have written either the proper name of the cluster or its number according to the Messier (M) catalogue or the general catalogue of clusters and nebulae (NGC). The segments of lines corresponding to globular clusters and parallel to the main sequence come to a halt in the lower part, as the distance of the globular cluster does not permit us to see the faint stars. In fact, they would also continue in the lower part of the diagram.

On examination of Figure 3, a series of very important conclusions can be drawn. First, the line of the main sequence for each galactic cluster diverges at a certain height to the right and it appears that the colder the cluster the lower the deviation on the diagram. The deviated main sequence breaks up almost at once. Moreover, in all galactic clusters groups of stars appear called red giants or supergiants (in young clusters, e.g. h, χ Perseus) which are 'torn off' the main sequence (small segments on the right-hand side of the diagram). The diagram of the old galactic cluster M67 resembles the corresponding diagram of a globular cluster, although its lower part coincides with the main sequence and the branch of the giants is lower. In globular clusters the main sequence is replaced by subdwarfs and the branch of the red giants is very clear and does

Table 2

Name of stellar population	Type of object belonging to the given population	Content of all elements besides hydrogen and helium, according to the number of atoms, %
Extreme population type 1	Stars of classes O, B. Very young galactic clusters and associations, cosmic dust, neutral interstellar hydrogen	4
Population type 1	Common stars of spectral classes from A to F. Galactic clusters. Red supergiants	3
Old population type 1	Stars of the main sequence and giants of spectral classes from G to K	2
Population type II	White dwarfs, numerous classes of variable stars	1
Extreme population type II	Globular clusters, subdwarfs	0.3

not come from the branch of the subdwarfs. It should be noted that the subdwarfs resemble stars of the main sequence, but at the same temperature their brightness is two or three times weaker. An important property of these stars is the very low abundance of heavy elements.

Galactic and globular clusters are part of the composition of our Galaxy. Thus, in our big stellar system we have at least two different types of, as we say, stellar population: a stellar population of the first type is composed of stars characteristic of galactic clusters and a stellar population of the second type is composed of stars characteristic of globular clusters. The majority of stars situated on the periphery of galaxies (the distance betwen the Sun and the centre of the Galaxy is of about 10 000 pc) belongs to the first type of population. Consequently, the colour of the central parts of galaxies is red and that of the peripheries blue, due to the presence of a great amount of bright and hot stars.

However, this division into two types of populations is too simplified. At present we distinguish five basic types of stellar population (Table 2). The following designations are often used: population of spherical subsystems (population II), intermediate and plane subsystems (population I). These designations are connected with the distribution of stars in the corresponding subsystems in space.

It is also very important that the abundance of all elements besides hydrogen and helium (see the last column of Table 2) increases in proportion with the decrease in the age type of the population. This conclusion was shown by a statistical analysis of stellar spectra belonging to different types of population. These properties of stellar populations are also explained by the theory of stellar evolution which will be considered later.

It should be noted that besides big stellar systems such as those of our Galaxy or the Andromeda Nebula, which are composed of all types of stellar population, there are many elliptic galaxies composed of only the type II population and galaxies of an irregular form in which the type I population prevails. It is evident that the stellar evolution is linked to the evolution of galaxies, but this question goes beyond the scope of this book. This is, for the time being, all we need to know for the study of the physics of stars.

2

A star—a sphere of gas

Let us start our study of stellar structure by resolving the following physical problem. Let us suppose we are given a huge sphere of gas in equilibrium with its own gravitational field. We must first determine the gas temperature in the central part of the sphere.

Let us consider the basis of the problem: according to Newton's law of universal gravitation any bodies attract each other with a force proportional to the product of their masses and inversely proportional to the square of the distance between their centres (the last condition is valid either for spherical bodies or for bodies of an arbitrary form but situated at large distances from one another)

$$F = f \frac{M_1 M_2}{r^2}$$

where F is the force of universal gravitation between two bodies of masses M_1 and M_2, r is the distance between them, and f is Newton's gravitation constant whose value is given on page 10.

Our gas sphere is composed of a great number of atoms, ions, and electrons (it is interesting to note that the number of particles in a star such as the Sun is expressed by unity with fifty-six zeros). All these particles attract each other according to the law of universal gravitation. It is true that for each pair of particles this force is very small, but the large number of particles makes the resulting gravitational force of our gas sphere sufficiently great. Under the influence of its gravitational force the gas sphere should contract—'collapse' towards the centre. However, in our case, according to a given condition, the sphere is in equilibrium. Therefore, a force should exist to counteract the gravity. This force is the gas pressure.

Indeed, if the gravitational force contracts the gas sphere, the gas pressure, inversely, tends to increase it. It is evident that the gas sphere will be in an equilibrium state only if the pressure of the gas in this sphere is equalized by the gravitational force. In other words, this problem can be expressed in the following way: the pressure of the gas close to the centre of the sphere should

17

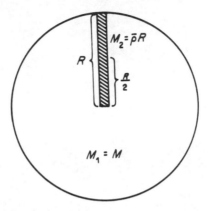

Figure 4. How to determine the pressure in the centre of stars

be equal to the weight of the column with a cross-section of 1 cm² and a height equal to the radius of the sphere (Figure 4). We remind the reader that the barometric pressure of the Earth's atmosphere can be calculated in an analogous way.

We shall now interpret this problem as a formula. It is evident that the weight of the hatched column in Figure 4 equals the force which attracts it to the centre of the sphere. In Newton's formula $M_1 = M$, the mass of the entire sphere, and M_2 is the mass of the column. If we designate the mean gas density in the column by $\bar{\rho}$, then $M_2 = \bar{\rho}R$, where R is the radius of the sphere (note that the cross-section of the column is 1 cm²). The distance between the centres of the sphere and the column is $r = R/2$.

Expressing the problem using this formula is not very precise. This is due to the facts that different parts of the hatched gas column are differently attracted. The upper parts are attracted by the whole sphere, but the lower parts, though situated in the centre, do not in fact experience the force of gravity. In order to explain this phenomenon we imagine ourselves to be at the Earth's centre. Where shall we fall? As a matter of fact the Earth will attract a body in its centre uniformly in all directions. As a result, being at the centre of the Earth we somehow lose our weight. In the same way, the gas in the centre of our sphere has no weight. Consequently, we cannot consider that our column will be equally attracted in all its parts towards the centre of the sphere. For a precise calculation, we must first know the density distribution inside the sphere and second use mathematical methods which go beyond the limits of a school level. As a matter of fact, we are not interested in a precise formula; we only need to know the order of magnitude. We must clearly explain the physical phenomenon and for this it is sufficient to use the simplest assumption that the selected column is uniformly attracted by the gas sphere.

Under these conditions the weight of the selected column (p) equals

$$(3) \qquad p = f \frac{M\bar{\rho}R}{(R/2)^2} = 4f \frac{\bar{\rho}M}{R}$$

If the sphere is in equilibrium this value should be numerically equal to the gas pressure close to its centre. To obtain an order of magnitude for this we must first calculate the pressure in the centre of a gas sphere with the dimension and mass of our Sun ($M = 2 \times 10^{33}$ g, $R = 7 \times 10^{10}$ cm, $\bar{\rho} = 1.4$ g/cm^3). We obtain $p = 10^{16}$ dyn/cm$^2 = 10^{10}$ atmospheres! Certainly, such a pressure cannot be obtained in our laboratories.

We know that the gas pressure increases as the temperature and density increase. We shall use the well-known law of gas composition or, as it is often called, the Clapeyron law. This law is as follows: the product of the volume of one gram-molecule of gas with its pressure divided by the absolute temperature is a universal gas constant A, numerically equal to 8.3×10^7 erg/mol grad. The formula of Clapeyron's law is very simple: $pV = AT$. It is, however, more convenient to present it in another form. We denote by μ the molecular weight of the stellar gas which is the same as one gram-molecule of the stellar matter. Then the stellar gas density $\rho = \mu/V$ and Clapeyron's formula can be written as follows:

$$(4) \quad p = \frac{A}{\mu} \rho T.$$

Let us come back to our gas sphere. The tremendous pressure at its centre can be obtained either with a large temperature or with a very high density (or with both conditions taken together). We know that the mean density of stars is small—about the same as the density of the matter on Earth. The density at the stellar centre is certainly greater than the mean density, but not much—perhaps a few times higher and at the worst a few scores of times higher—and in any case it cannot increase the pressure up to milliards of atmospheres.

We have touched upon this question of central stellar densities so in order to avoid further misunderstanding we shall make two remarks: first, although the densities of white dwarfs are great, the matter in their innermost layers is not a normal gas and therefore what has been said above (and in particular Clapeyron's law) cannot be applied; second, the mean densities of giants, and in particular of supergiants, are very small but they have a complex inner structure and therefore the above reasoning is not valid in their case. The white dwarfs and red giants will not enter into our considerations for the time being and we shall come back to them later.

There is one hypothesis left. The great pressure indispensible for the retention of the gas sphere in the equilibrium state can be assured only by very high temperatures. In order to determine this we equate the gas pressure with Clapeyron's formula for the centre of the sphere to the weight of the column. We obtain

$$p_c = \frac{A}{\mu} \rho_c T_c = 4f \frac{\bar{\rho} M}{R}$$

where ρ_c and T_c are the density and temperature in the centre of the sphere. The mean of the central densities enters into this formula as well. If the gas spheres have the same structure, but a different mass and consequently a different radius, then both these values (ρ_c and $\bar\rho$) are proportional to one another. How many times is ρ_c bigger than $\bar\rho$? This depends on the concrete structure of the gas sphere and can be answered only after a detailed study. As we do not treat the precise theory here but are only interested in the physical aspect of the problem, it is sufficient to take, for example, $\rho_c = 4\bar\rho$. The coefficient 4 proves to be correct later (with this value the temperature calculations from formula (5) are in good agreement with more precise calculations). Then for the central temperature of the gas sphere we have

$$(5) \qquad T_c = \frac{\mu f M}{AR}$$

This formula determines the temperature at the centre of the star according to its mass, radius, and molecular weight of the stellar matter. Its importance for the theory of inner stellar structure will be proved many times. We also must mention that even though it has been deduced quite approximately, after a few essential simplifications, it is, in fact, precise enough. The central temperatures calculated with this formula for stars of the main sequence differ at the most from the effective ones by not more than 10–20 per cent. or less.

The masses and radii of stars are known, as well as the constants A and f. The molecular weight is left to be determined. The expression 'molecular weight' defines the mass of the matter composed of a fully determined number of particles equal to the so-called Avogadro number 6×10^{23}. For example, the molecular weight of atomic hydrogen (it is also called the atomic weight) is the mass of 6×10^{23} hydrogen atoms numerically equal to unity, since the mass of one hydrogen atom is 1.67×10^{-24} g. The molecular weight of gas composed of hydrogen molecules equals the mass of 6×10^{23} hydrogen molecules and therefore equals two, etc. The molecular weight of a mixture of gases (e.g. air) is determined in a similar way: take a general number of particles equal to the Avogadro number (each component of the mixture should be taken in corresponding proportion) and then find the sum of the masses. This is the way we proceed to determine the molecular weight of stellar matter. The stellar matter is composed of atoms and ions of diverse elements and therefore we should first of all determine its chemical composition.

We shall determine here the molecular weight of the stellar matter in the interior of the stars, i.e. where the gas temperature reaches millions of degrees. Of course, at such temperatures molecules cannot exist. Moreover, at this temperature all atoms must be ionized. The most 'stable' atoms of inert gases are ionized, i.e. they lose one or several electrons, at a few tens of thousands of degrees. Thus, the material in stellar interiors is a mixture of electrons and the 'residues' of atoms (atomic nuclei with 'interior electrons' that have survived or else nuclei altogether 'bared of electrons'). First of all we have to find out how many electrons have been torn off the atoms, i.e. to what degree the stellar matter has ionized.

Let us give some numbers. At temperatures of millions of degrees the mean energy of the particle is about 2.4×10^{-10} erg. On the other hand, the energy of attraction of the electron towards the proton (coupling energy) in a hydrogen atom equals approximately 2.2×10^{-11} erg, which means that it is about ten times smaller. It is clear that all hydrogen atoms in the stellar matter are split into electrons and protons at a temperature of more than a million degrees, because each collision of a hydrogen atom with another particle causes its disintegration. The situation is the same for helium atoms. Although the coupling energy binding two electrons to the nucleus is much greater than the energy of coupling in a hydrogen atom, it is still smaller than the energy of separate particles; the great majority of helium atoms in stellar matter at temperatures exceeding a million degrees disintegrates into electrons (two for each atom) and alpha particles.

The energies of electrons in atoms of other elements are very different. The fact is that in a complex atom the electrons revolve around a nucleus, not on one orbit but on different orbits situated at different distances from the nucleus. The inner electrons are more strongly bound to the nucleus than electrons in hydrogen or helium atoms and inversely the outer electrons are more weakly bound. In this case the heavier the nucleus and the more electrons in the atom, the greater the binding energy of the inner electrons, whereas the binding energy of the outer electrons changes very little; it depends on the chemical properties of the elements. For this reason in the stellar matter the outer electrons are torn off all atoms, whereas the inner electrons can remain with the nucleus. It is evident that the heavier the atom, the greater the charge of its nucleus and the more inner electrons it keeps. At a temperature of about a million degrees the nuclei of oxygen, nitrogen, and carbon keep two inner electrons and the heavier elements also keep the inner electrons at such temperatures. However, at a temperature of about ten million degrees these nuclei can no longer keep even the 'most strongly bound' electrons. At a temperature of ten million degrees they are almost entirely ionized and the most abundant elements, after hydrogen and helium, are oxygen, nitrogen, and carbon.

We can therefore assume that at a temperature of about ten million degrees the stellar matter is no longer composed of atoms, but of electrons, protons, alpha particles, and 'bare' nuclei of other elements. At lower temperatures the nuclei of all elements, besides hydrogen and helium, can still keep some inner electrons. Let us note that if a nucleus has entirely lost its electrons it does not always stay 'bare'. From time to time the nucleus of an atom captures an electron and keeps it for a while before losing it again. Protons and alpha particles can also capture electrons but lose them very rapidly.

While calculating the molecular weight of the stellar matter we must consider one important property of atomic weights, or the chemical elements situated at the beginning of Mendeleev's table. It is known that the elements of the first three periods in Mendeleev's table (except hydrogen) have an atomic weight approximately twice as big as their atomic number. On the other hand, since the atomic number determines the charge of the nucleus and, subsequently, also the number of electrons in the atom, we can say that for these elements the atomic weight is about twice as big as the number of electrons in the atom. If we consider all particles (electrons and nuclei) this relation will hardly be changed. Thus, for example, if an oxygen atom is fully ionized, nine particles are formed (eight electrons and one nucleus). The atomic weight of oxygen is 16; consequently, one particle represents on the average $16/9 = 1.8$ of an atomic weight unit, i.e. a value close to two. It is evident that at the complete ionization of any element in the first three periods of Mendeleev's table (except hydrogen and helium) the mean atomic weight for one particle will also approach two.

After the complete ionization of helium three particles are formed. Dividing the atomic weight of helium ($\mu = 4$) by three, we see that on the average for one particle there are 4/3 of an atomic weight unit. Finally, after complete ionization of an hydrogen atom two particles are formed (one electron and one proton) and each particle has 1/2 of an atomic weight unit.

Let us now perform a simple calculation to find the molecular weight of the stellar matter. Remember that 6×10^{23} particles of the stellar matter (electrons and atomic nuclei) have a mass equal to μ. In one gram-molecule of stellar matter are μX grams of hydrogen. Dividing this value by 1.67×10^{-24}g (the mass of one hydrogen atom) we obtain the amount of hydrogen atoms in one gram-molecule. Since each hydrogen atom is divided through ionization into two particles (a proton and an electron) the general amount of particles in one gram-molecule is $2 \mu X/(1.67 \times 10^{-24})$. In the same way, considering helium and taking into account the fact that the mass of a helium atom is $4 \times 1.67 \times 10^{-24}$ g we get a number of particles (electrons and helium nuclei) for one gram-molecule of stellar matter, by ionization of helium, which is $3\mu Y/(4 \times 1.67 \times 10^{-24})$. The sum of the masses of all other atoms under the same conditions is μZ. Since one particle has an atomic weight approaching two, that is $2 \times 1.67 \times 10^{-24}$, the number of particles generated by ionization of all elements, except hydrogen and helium, equals $\mu Z/(2 \times 1.67 \times 10^{-24})$. The total number of particles in one gram-molecule equals the Avogadro number. Subsequently,

$$\frac{2\mu X}{1.67 \times 10^{-24}} + \frac{3\mu Y}{4 \times 1.67 \times 10^{-24}} + \frac{\mu Z}{2 \times 1.67 \times 10^{-24}} = 6 \times 10^{23}.$$

Now we can find the final formula for the calculation of the molecular weight of the stellar matter:

$$(6) \quad \mu = \frac{1}{2X + \frac{3}{4}Y + \frac{1}{2}Z}$$

For example, if in the stellar matter there were neither hydrogen nor helium ($X = Y = 0$, $Z = 1$), then $\mu = 2$. Inversely, stellar matter composed of pure hydrogen ($X = 1$, $Y = Z = 0$) has a molecular weight of 0.5. Thus, the molecular weight of the stellar matter should be within the limits from 0.5 to 2.

Table 3

Star	Temperature in millions of degrees	Star	Temperature in millions of degrees
ζ Arietis	43	Procyon	11
Vega	13	ξ Boötis	12
Sirius	15	η Cassiopeiae	9

The percentage of 'heavy' elements in stars is different for stars of diverse populations (see Table 1), but does not exceed 4 per cent ($Z \leqslant 0.04$). For this reason, the molecular weight depends mainly on the relative contents of hydrogen and helium. In this way we realize that in the evolution process of a star its hydrogen is converted into helium and subsequently, as time passes, the molecular weight of its stellar matter also changes. In a helium star ($X = 0$, $Y = 1$, $Z = 0$) $\mu = 1.3$. Therefore, in fact, the molecular weight of the stellar matter varies within smaller limits—from 0.5 to 1.3. In particular, for stars in the middle part of the main sequence, which also contains our Sun ($X = 0.71$, $Y = 0.27$, $Z = 0.02$), $\mu = 0.6$.

We have now learnt how to calculate the molecular weight of the stellar matter and have all the data necessary to calculate the central temperature of the main sequence stars. For this it is convenient to transform formula (5) by expressing M and R as ratios to magnitudes of the Sun, i.e. we express M in units of 2×10^{33} ($M/M_\odot$) and R in units of 7×10^{10} ($R/R_\odot$). We also assume that the hydrogen and helium abundance in stars of the main sequence is approximately the same as in the Sun and we take $\mu = 0.6$. We then obtain a very simple formula to calculate the central temperature of stars:

$$(7) \quad T_c = 14 \left(\frac{M}{M_\odot} \right) \left(\frac{R_\odot}{R} \right) \quad \text{million degrees}$$

According to this formula the temperature in the centre of the Sun equals 14 million degrees (a more precise calculation yields almost the same result). The central temperatures of other stars of the main sequence given in Table 1 are shown in Table 3. As expected, in brilliant and hot stars the central temperature is very high and reaches a few tens of millions of degrees; in cooler stars it reaches about ten million degrees.

The difficult problem of the definition of the temperature in the centre of stars has been resolved quite simply on the basis of two well-known physical laws: the law of universal gravitation and the law of the state of a gas. It is true that we made many simplifications, but a precise calculation is also based on the same physical laws, using only more perfect mathematical techniques, since astronomers need to know the central temperatures with greater precision.

The next stage in the study of the physics of stars will be limited to the study of the processes which led to the presence of very high temperatures in the stellar interiors and comparatively low temperatures on their surfaces.

3

Energy transfer in stars

We already know that a star in an equilibrium state must have an inner temperature reaching millions of degrees. It is evident that there must be sources of energy heating the stellar matter. In fact, it is well known that due to a difference of temperature heat is transmitted from hotter bodies to cooler ones. The same happens in stars, where the heat is transmitted from the very hot central parts of the star to the comparatively cool surface from where it radiates into space. The sources of stellar energy balancing these losses will be examined in the following chapter but here we shall study the process of heat transfer in stars.

A question arises: is it important for the theory of stellar structure to study in detail the energy radiating from the interior of stars towards the exterior? It seems to be very important. First, this energy flux determines the structure of almost all parts of the star; second, the luminosity of the star—one of its fundamental parameters—is simply energy flux 'leaving' the interior of the star towards its surface. For this reason it is very important for us to know the physical processes of energy transfer in the star and to know how to calculate this energy flux in order to be able to compare the theory with observations.

In physics one studies diverse possibilities of energy transfer; we are acquainted with these phenomena in everyday life although we do not always notice them. For example, the heat from a hot body can be transmitted to a cooler one by means of thermal conductivity. If you strongly heat one end of an iron bar, the other end will be heated too—the heat from the heated end diffuses along the bar due to the fact that there are free electrons in the metal which transport the thermal energy. This kind of energy transfer, called thermal conductivity, has no significance in ordinary stars (but is important for white dwarfs).

Another means of heat transmission is convection. The air adjacent to the heated body is also heated and rises carrying away the heat. This warm air is replaced by cool air which is heated in its turn and thus also carries away the heat. The heated air coming into contact with cool bodies transmits the heat—in this way the energy transfer is realized by convection. The convective heat transfer in stellar interiors is possible and is in fact frequently observed. We shall study convection later.

The most important means of energy transfer in stellar interiors is radiation. This type of energy transfer is often observed in everyday life. It is well known that on approaching a heated body (a stove, a heated piece of iron, etc.) we immediately feel the heat. This heat comes only from the heated object so evidently this heated object emits rays which transport energy (and from which we can protect ourselves with a screen). Usually these rays are infrared and invisible, but if we heat a metal to red heat then the thermal radiation becomes visible—it is a light emitted by the heated body.

We have already met this phenomenon in stars while studying the fundamental parameters. In fact, the radiation of energy from a star (e.g. the Sun) and the absorption of this energy by a cool body (e.g. the Earth) are also processes of heat transfer by radiation. We have learnt that the colour and the amount of radiated energy are determined in the first place by the temperature of the heated body. We do not need to know the colour of radiation in stellar interiors, as we do not see it, and the amount of radiated energy, as we have seen, is proportional to the forth degree of the temperature (the Stefan–Boltzmann radiation law). For this reason the ability to radiate increases very rapidly in proportion to the screening of the stellar interior. For example, each cubic centimetre of the stellar matter in the centre of the Sun radiates about 10^{26} erg in one second. If the Sun were transparent in all directions, this energy would burst out and reduce everything on Earth to ashes. However, as the stellar matter is not transparent, the luminous energy must 'filter' through the mass of the star. We shall illustrate this process with examples.

Imagine that we have a body heated to a quite high temperature and that there are two or three iron screens, one behind the other, between the body and ourselves. Will we feel the heat coming from the body? Yes, although very faintly. In fact, the heat rays falling on the first screen are absorbed and heat it. This screen then starts to emit heat. Part of it is emitted to the heated body and part of it to the second screen, which, in its turn, absorbs it and consequently is heated. The second screen starts to emit heat, part of which falls on the third screen. Thus, although we protected ourselves by opaque screens, part of the heat from the hot body will reach us. In the same way, the thermal energy in stars, emitted by the central parts, is absorbed by the upper layers and is then emitted again, thus 'filtering' towards the surface of the star.

Let us try to represent this process by a formula with certain reserves; its rigorous conclusion needs the use of higher mathematics and therefore we will not give it here. An approximate estimation of this process is very intelligible. Let us go back to Figure 4 and study the flux of thermal energy through the shaded columns. The upper end emits energy, as it is easy to imagine, equal to the luminosity of the star, which is distributed over its surface, that is $L/(4\pi R^2)$. This is what reaches the surface from the energy source, which we can assume to be located in the centre, through the entire column of the stellar matter which acts like the 'screens' of the preceding example. If we extracted this column from the star, then its base would emit an energy which could be calculated using the Stefan–Boltzmann law with the central temperature of the star. Thus, at the base of the column, energy is emitted equal to

$$\sigma T_c^4 = \sigma \left(\frac{\mu f M}{AR} \right)^4$$

Remember that σ is the Stefan–Boltzmann constant (see Page 8). Moreover, we used here the formula for the central temperature of the star, obtained in the preceding chapter. If we invert our column (we 'introduce the screens') we notice that the emission will weaken. In our example with the screens covering the heated body, the larger the screen, i.e. the greater the total opacity, the greater is the decrease in the thermal flux. The same thing happens here; the greater the opacity of the column, the stronger the decrease of the thermal energy flux.

Now we have an important problem: how do we explain what the opacity of the stellar matter is? Let us first consider transparency in general. Any body is called transparent if a ray of light passes through it in a straight line without difficulty. In opaque matter the rays of light can no longer pass through in a straight line; they are either completely absorbed and their light energy converted into heat or they are diffused in different directions—part of it returns, part is deviated through different angles, and only a small part passes through the body. It is evident that we can see nothing or almost nothing through such a body and therefore call it opaque. The opacity of a body depends on its thickness (any matter if reduced to a thin film will become transparent) and on the density; in general, the denser the matter, the less transparent (the hard, dense matters are usually opaque, fluids are often transparent, and gases are almost always transparent), although on Earth there are many exceptions to this principle. Finally the transparency depends on chemical and physical properties of the matter.

The stellar material is a gas and consequently should be transparent. However, as a matter of fact, in stellar interiors this matter is very densely compressed and therefore becomes opaque. The transparency of the stellar matter in the centre of the Sun can be compared to the transparency of wood. Radiation can, however, still 'filter' through it.

The general opacity of our entire column should be, according to what has been said above, proportional to the mean density and height of the column. The factor of proportionality, which we designate $\varkappa$ (of a dimension $\mathrm{cm}^{-1}\,\mathrm{g}^{-1}$), is called the 'opacity'.

The radiation leaving the base of the column while filtering through the entire mass will become $3\varkappa\rho R$ times weaker (the factor 3 takes into account the inhomogeneous distribution of the stellar matter over the radius). Subsequently, the amount of energy reaching the surface is

$$\frac{L}{4\pi R^2} = \frac{\sigma T_c^4}{3\varkappa\bar\rho R} = \frac{\sigma}{3\bar\varkappa\rho R} \left(\frac{\mu f M}{AR} \right)^4$$

We introduce $\bar{\rho} = 3M/(4\pi R^3)$ and obtain the formula

$$(8) \quad L = \frac{16\pi^2 \sigma f^4}{9A^4} \frac{\mu^4}{\varkappa} M^3 .$$

Since formula (8) is very important for the theory of inner stellar structure we shall once more derive the same formula.

It is known that the light exerts a pressure on the absorbing or reflecting medium. The magnitude of the radiation pressure can be calculated with the formula $P_{rad} = (4\sigma/3c)T^4$, where c is the velocity of light. It is evident that the radiation pressure in the centre of a star equals $4\sigma T_c^4/3c$. This pressure should be equal to the pressure of the flux of radiation absorbed on the entire way to the surface. The amount of light absorbed on 1 cm of the path while the radiation filters through 1 cm^2 at a distance r from the centre equals $\varkappa\rho L/(4\pi r^2)$. The total amount of absorbed light in the entire path from the centre of the star to the outside in a column with a cross-section of 1 cm^2 equals $\varkappa\bar{\rho}L/(\pi R^2)$ (we put $\bar{\rho}$ for ρ and $R/2$ for r). The absorbed light exerts a pressure on the stellar matter in this column. In order to determine its magnitude we recall that in the study of the light pressure we discovered that the momentum (or pressure) of one light ray is always c times smaller than its energy. For this reason the total momentum of radiation absorbed during the filtering from the centre of the star to its surface equals $\varkappa\bar{\rho}L(\pi R^2 c)$. This magnitude should also be compared to the radiation pressure in the centre, which leads to formula (8).

Formula (8) is called the theoretical mass–luminosity relationship. Before we compare it to observations we shall examine its components. The first one, $16\pi^2\sigma f^4/(9A^4)$ is a constant composed of physical constants and is equal to 4.2×10^{-64}. The molecular weight μ depends on the chemical composition and we already know this dependency (formula 6). The opacity $\varkappa$ also depends on the chemical composition of the stellar matter. Let us study this dependency.

It is easy to understand that protons, alpha particles, and 'bare' nuclei of other elements have very little influence on transparency since they do not have nearby electrons capable of absorbing light during the transition between orbits. It is true that from time to time these nuclei capture electrons for a short time and then for a short period are capable of absorbing light. Moreover, if a free electron passes close to an atomic nucleus, absorption of light is also possible, but the role of these processes is small. In principle, the opacity of the stellar material is due to the presence of atomic residues of heavy elements which can only keep their inner electrons in the conditions which exist in stellar interiors. Therefore the opacity coefficient should be proportional to the percentage amount of heavy elements Z.

Yet this is not all. As the temperature increases, even heavier nuclei of oxygen, neon, etc., start to lose their inner electrons. These then stop absorbing light and, as a result, the opacity of the stellar material decreases. As the density of the stellar material increases the opacity increases since a closer disposition of the particles increases the probability of the capture of free electrons by the 'bare' atoms. Thus the opacity increases as the density

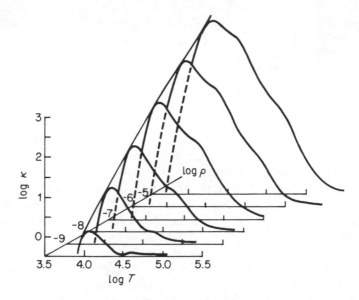

Figure 5. Dependence of the opacity on density and temperature. The curves on the graph indicate the variation of the opacity with the change in temperature at a given value of density

increases, the temperature decreases, and the amount of heavy elements increases. It is true that close to the stellar surface the opacity coefficient increases with temperature, since at large T there are more excited atoms capable of absorbing light in the visibe spectral region; e.g. in cool stars of population I (a large abundance of helium and heavy elements) $\varkappa \sim \rho^{0.7} T^{6.3}$. In cool stars of type II population (less helium and heavy elements) the absorption coefficient depends even more strongly on temperature: $\varkappa \sim \rho^{0.6} T^{10.4}$. Incidentally, this high degree of dependence of $\varkappa$ on T also leads to the fact that the surface temperature of such stars cannot change very much ($T \approx 3$–4 thousand degrees).

On the other hand, on the surface of very hot stars ($T > 20\,000\,\text{K}$) the absorption coefficient is almost constant ($\varkappa \approx 0.2\ \text{cm}^2/\text{g}$). Here the opacity can be explained by the diffusion of light from the free electrons.

For more profound stellar layers Kramer's rule $\varkappa \sim \rho/T^{7/2}$ is frequently used. Of course, in modern calculations this formula is not precise enough so more detailed tables must be used.

Figure 5 shows a graph of the evolution track of the dependence of the opacity on temperature and density. The left-hand slope corresponds to low temperatures (i.e. atmospheres of cool stars) while the right-hand slope is an example of Kramer's rule. In a region of even higher temperatures the opacity tends towards a constant value; this is the region of Thomson's diffusion.

As we go deeper into the star ρ and T increase and therefore we can expect that $\varkappa$ varies slowly. Precise calculations have shown that in fact, although the temperature increases about one thousand times on the way from the surface to the centre, the opacity changes not more than ten times. For this reason the error is not large in the arguments given above, assuming that $\varkappa$ is the same over the entire length of the column. From one star to another $\varkappa$ will change very little. In hot and brilliant stars $\varkappa \approx 0.7 - 3\ \mathrm{cm}^2/\mathrm{g}$ while in cooler stars of the main sequence $\varkappa \approx 8 - 25\ \mathrm{cm}^2/\mathrm{g}$. This means that a layer of stellar gas only a few millimetres thick, or even more, is already opaque.

Now let us return to the theoretical mass–luminosity relationship. In order to compare it with observations we transform formula (8) in a way analogous to the transformation of formula (5) into formula (7). We obtain

$$(9) \quad \frac{L}{L_\odot} = 860\,\frac{\mu^4}{\varkappa}\left(\frac{M}{M_\odot}\right)^3.$$

With this formula we calculate the theoretical luminosity of the Sun. The opacity is taken to be $20\ \mathrm{cm}^2/\mathrm{g}$. Introducing $\mu = 0.6$ and $M = M_\odot$ into (9) we find $L_{\mathrm{theor}} = 5.6 L_\odot$.

To sum up. Considering the process of energy transmission from the stellar interior to the outer surface we obtain a formula which relates the luminosity of a star to its mass. The observations in general satisfy this relationship (see Figure 3), but the theoretical luminosity of the Sun appears to be five or six times greater than the observed one. How can this be so? It is possible that we have a false conception of the physical processes which led to the theoretical deduction of the mass–luminosity relation, or else the difference is simply due to the approximations in our calculations. We calculated the central temperature of the star with formula (5) which, as we already noted, can contain an error of 10–20 per cent. Since the luminosity is proportional to T_c^4. the decrease of T_c by 20 per cent. will diminish the theoretical value of luminosity by a factor of 2. Moreover, the assumption of a constant value $\varkappa$ along the stellar radius also gives an error of a few times. Thus, the difference between the theoretical and observational luminosity magnitudes can be entirely due to the approximations in our calculations. There is no need to complain; a precise calculation needs the solution of a quite complex system of differential equations which are nowadays usually performed on electronic computers. We should rather consider, as a quality, the fact that such a simple and approximate calculation has yielded a comparatively good agreement with the theory.

It is important to understand correctly the physical phenomena which are at the base of the observed mass–luminosity relationship. Figure 2 shows clearly that the luminosity increases as the mass increases. what is this relation? If there were, in all stars, the same values for the molecular weight and the opacity, then there should be $L \sim M^3$. An analysis of Figure 2 shows that effectively in massive $(M > 3M_\odot)$ and hot stars $L \sim M^3$. This is explained by the fact that in these stars there is total ionization almost everywhere (constant μ) and almost the same value of $\varkappa$. In fact, we have already noted that in hot stars the opacity is linked to the diffusion of light from free electrons and that each electron screens the light from one section.

In stars with smaller mass $(M < 3M_\odot)$ the opacity is more sensitive to temperature and density variations and therefore $\varkappa$ changes from one star to another. Indeed, from Figure 2 is it seen that the variation of luminosity with mass is a little sharper (approximately as $L \sim M^{2.5}$). Thus Figure 2 corresponds perfectly to equation (9) if we also take into account the variations of the factor $\mu^4/\varkappa$ in the transition from one star to another.

The theory, therefore, not only permits us to calculate the luminosity magnitude but also to describe correctly the dependence on its mass. This can serve as the best confirmation of the correctness of the theory. The fundamental conclusion is: the luminosity of the star is simply determined by its mass, since this rule is based on one physical phenomenon, the transfer of energy from the stellar interior towards the outside by radiation, while the 'capacity' of the stellar material is in the end determined only by the mass of the star. The sources of the stellar energy must somehow adapt to the stellar opacity. One of the two fundamental relations between the basic parameters of stars can be explained in this simple and natural way.

We shall often come back to the different consequences of the mass–luminosity relationship. Let us simply note here that this relation was first deduced theoretically by the English astrophysicist Eddington and later confirmed by observations.

The phenomenon of convection was briefly mentioned in the beginning of this chapter. We note once again that in stellar conditions convection is very frequently observed. In contrast to the energy transfer by radiation, where the amount of transmitted thermal energy is determined by the opacity of the matter and is therefore limited (this limitation of the radiant flux in the star also appears as a deduction from the mass–luminosity relationship), the convective energy flux is not limited. In fact, when observing heated water in a kettle one can see that the more we heat the water the faster the convective motions and consequently the faster the transmission of heat from the heated part of the kettle to the whole mass of water.

In stars the transfer of thermal energy is realized mostly by radiation, but there are also layers of stellar material where the thermal energy is transmitted towards higher layers by convection. Convection appears where the energy transfer by radiation seems not to be sufficient: either the stellar matter became too opaque or the temperature decrease is so strong that the energy, so to speak, 'overflows'. However, the fact that a layer with convective energy transfer exists in stars does not upset the correctness of the theoretical deduction of the mass–luminosity relation.

If there is only one layer with radiative energy transfer, then it is indeed the opacity of this layer which determines the 'transmission' capacity of the entire mass of the star and subsequently also its luminosity. Only the coefficient in the equation for the mass–luminosity relation can change (this is, by the way, one more possible cause of the discrepancy between the theoretical and observational values of the luminosity of the Sun). Among the majority of the main sequence stars there are no entirely convective stars; therefore the mass–luminosity relationship has a universal character for these stars.

Nevertheless, in nature there probably also exist entirely convective stars. Apparently, in stars of very small mass, convection diffuses over the whole star. In newly formed stars, not yet in a steady state (they are called protostars), the energy is also transferred by convection. However, in these objects the convective transfer is also replaced in the surface layers by radiative transfer.

Since convective transfer plays an important part in stars—particularly in non-steady-state stars—we shall study it in detail. Let us consider a case with a rapid temperature increase as we penetrate deeper into the star.

If at a great depth in a given mass of gas the temperature has increased by chance, then this mass will start expanding. Its density will decrease, it will now be lighter than its surrounding stellar matter, and it will therefore rise. As it rises, it will pass through always cooler outer layers of the star, where the pressure is also respectively smaller; therefore our mass of gas will expand even more and cool. If the temperature decrease in the stellar interior is sufficiently big, then the rising mass of gas appears to be, as previously, hotter than the surrounding stellar matter and its rise will continue. In the end, however, this mass of gas will transmit its heat to the surrounding stellar matter at an even greater height and will then come to a stop. Therefore, a certain mass of gas having initially a certain excess of thermal energy transfers all its thermal energy supply from the lower layers to the cooler upper layers, thus realizing the convective transmission of thermal energy. In the place of the rising mass of heated gas, cooler masses of gas come down from the higher levels, which in their turn are heated and the same process starts all over again. It is easy to understand how convection can transfer an unlimited amount of thermal energy: the more energy that has to be transferred, the greater the mass taking part in the convection and the greater the velocity of the rising heated mass (and according to this the greater the velocity of the descending cooler mass).

In the convective transfer of thermal energy in stellar interiors the energy flux is no longer determined by a simple formula. According to what has been said above, we can consider that convection transfers only the amount of thermal energy that is emitted. On the other hand, we can theoretically derive, from the convection graph, the relation between the gas density and its temperature, and vice versa. In fact, one can consider that the rising mass of gas expands adiabatically, since during its rise it keeps its initial supply of thermal energy until it comes to a stop. It is known that for an adiabatic variation of the state of a gas its pressure is proportional to the density of degree γ (where γ is the thermal capacity at a constant volume). The stellar matter is a gas of single atoms (in fact, the electrons, protons, and 'bare' nuclei also have only three degrees of freedom of kinetic motion) and therefore it has $\gamma = 5/3$. Consequently, for an adiabatic change of the stellar matter the pressure P is proportional to $\rho^{5/3}$. We designate the pressure and the density at the beginning of the rise of the convective mass of gas by the index 2. Then, since during the rise the adiabatic conditions are satisfied, we obtain the relation

$$(10) \quad \frac{P_2}{P_1} = \left(\frac{\rho^2}{\rho_1} \right)^{5/3}.$$

We note that in a rising mass of gas the temperature is somewhat higher than the temperature of the surrounding matter, but only very little, probably by only one degree, which is very small compared to stellar temperatures of millions of degrees. From this follows that for a convective energy transfer the pressure and density of the surrounding matter satisfy the condition $P/\rho^{5/3} = a$ constant. This condition also determines the structure of the stellar layers with a convective energy transfer. We shall give two more values characterizing the stellar convection: the velocity of the rising mass of gas is of the order of 30 m/s and the time of the rise is about 20 days.

As we are now acquainted with the energy transfer in stars and have obtained at the same time the theoretical explanation of the mass–luminosity relationship, we can now study the sources of stellar energy. But first we need a historical background.

Astronomy is an ancient science, possibly the oldest science among all others. Nevertheless, the study of stellar structure and evolution is one of its youngest chapters. Four hundred years ago Giordano Bruno taught that all stars are celestial bodies resembling our Sun, but two hundred years before that people thought that the Sun was a hard sphere, the same as planets, but covered by hot clouds. At present it is hard to tell who was the first to conceive that stars, among them the Sun, are spheres of gas. The first calculations of the structure of these gas spheres were performed by D. Len and A. Ritter in 1869–1878. The energy transfer with the help of convection in the inner regions of stars was studied by Kelvin (England, 1887), but most important for the theory of stellar structure is, as we know, the energy transfer by radiation. The notion of radiative energy transfer in stars was first given by a Polish scientist Bielobrzecki (1913), but it was further developed in the works of the well-known English astronomer A. Eddington. Indeed, Eddington considered the importance of energy transfer in stars and obtained the mass–luminosity relationship theoretically, which he then compared to observations. He then elaborated the first standard stellar model. The year of Eddington's *Inner Structure of Stars* publication, 1962, is considered as the moment of origin of this theory.

4

Thermonuclear sources of stellar energy

In nature there exists a universal law of energy conservation: energy cannot be formed from nothing, it only turns from one form into another. For example, from the combustion of a chemical fuel the energy of attraction of the atoms in the molecules of this fuel turns into heat. This is chemical energy. It is clear that stars cannot dwell on chemical energy—at a temperature of millions of degrees any molecules immediately disintegrate.

Maybe the star simply releases its stock of thermal energy. Let us consider this process. If the energy released by the star is taken from its thermal energy, then with time the temperature in the stellar interior should decrease. However, this would disturb the equilibrium of the star, since at a lower temperature the gas pressure is no longer able to counteract the gravitational force of the star. The star would then contract and in a way collapse towards the centre. At any decrease of pressure the potential energy turns into kinetic and thermal energy (e.g. if a body falls through air it is heated by friction). Consequently, at the contraction of a star gravitational energy is also released (potential energy of the gravitational force of the star). Part of this energy gives an increase in the temperature of the stellar interior, preventing in this way a too rapid contraction, and part of the energy is released into space. Furthermore, if we consider this process in detail it will appear that the temperature of the centre of the star, deprived of the sources of stellar energy, does not decrease but on the contrary increases, i.e. the star does not cool but is heated thanks to the release of potential gravitational energy. This can be confirmed by formula (5) which is also valid for a contracting star only if the contraction is not very rapid. With a decrease in the radius at constant mass the central temperature increases.

How long will the stock of potential energy in a star last? This can be easily calculated. It is known that the potential energy of a body in a gravitational field equals Mgh, where M is the mass of the body, g the acccleration of gravity, and h the height of the body above the level taken as the beginning of the reading. Let us try to apply this formula to our case. The whole star contracts in a gravitational field. Consequently, M is the mass of the star. To accelerate the gravity we take its value on the surface: $g = fM/R^2$. In the stellar interior its value is approximately the same. Finally, calculating the entire supply of potential energy we must evaluate the total contraction of the star to a small dimension and therefore the length of 'collapse' of the star can be equal to its radius: $h = R$. Subsequently, the total supply of potential energy of the proper gravity of the star equals fM^2/R. A precise calculation (applying higher mathematics) yields the same formula, but only with a numerical factor close to unity.

During the contraction of the star approximately half of the energy is used to heat the star and the other half is released into space. Since the luminosity of the star depends simply on its mass, during contraction with a constant mass the luminosity will not vary either. Consequently, the duration of contraction of a star, or the time for which there will be enough potential energy, is

$$(11) \quad t = \frac{fM^2}{2RL}$$

For the Sun $t = 5 \times 10^{14}$ s $= 1.6$ millions of years. This is a very small value. The Earth, and therefore also the Sun, has existed for at least milliards of years. Thus, in the Sun, as well as in other stars of the main sequence, there should exist another source of energy, which we shall consider later. Nevertheless, there is a group of stars 'living' on account of the contraction energy. These are recently formed stars.

Formula (11) shows that the stage of stellar formation should be very short. Note that the release of stellar energy on account of contraction in a proper gravitation field is called 'a gravitational energy source'.

The problem of stellar energy sources has been of interest to astronomers for a long time. Many hypotheses have been framed, among them the hypothesis of nuclear reactions was discussed (in the twenties). For a long time these were only general reflections and only in 1938–1939 were the American physicist G. Bethe and others able to calculate theoretically just which concrete nuclear reactions are sources of stellar energy.

Nuclear reactions in general, and thermonuclear reactions in particular, represent a very large chapter in modern physics and also have important technical applications. Unfortunately, it is not possible to consider this problem fully here, the reader will have to turn to other sources which specifically deal with nuclear reactions in more detail. Here we will limit ourselves to a brief outline.

You will remember that the nucleus of an atom of a chemical element is composed of protons and neutrons (except the nucleus of a hydrogen atom, which has only one proton). Protons and neutrons are linked in the nucleus by very strong forces called nuclear forces. The nature of these forces is not yet known but it is important to know that such forces exist and that the energy of this coupling is very large. By the term 'coupling energy' we mean the work which must be spent to destroy the nucleus and pull apart its particles to large distances. The magnitude of these forces can be judged by the fact that they must be able to surmount the repulsive forces between protons, since all protons have the same positive charges. Note also that the nuclear forces are such that in most cases of not too heavy stable nuclei the number of protons and neutrons is about the same. This rule is violated only in heavier nuclei. It is evident that if one more proton or neutron arrives in the nucleus of an atom, a new nucleus is formed that can be stable as well as unstable, depending on the relation between the protons and neutrons in the initial nucleus. Moreover, in this case energy is usually emitted since the 'newly arrived' particle is linked by the nuclear forces to the other particles and the excess energy must be released. This excess energy can either be radiated in the form of gamma rays or carried off by a particle ejected from the nucleus (by a proton, neutron, or even an electron or positron if the so-called beta-decay takes place in the nucleus). This is a nuclear reaction. Thus, in a nuclear reaction a new nucleus is formed and energy is released.

Nuclear reactions may also be more complex, e.g. when not one particle (proton or neutron) but the nucleus of another element arrives in the original nucleus. Very frequent nuclear reactions take place with alpha particles and helium nuclei. There exist reactions of an altogether different type—division or fission reactions—where a more complex nucleus is divided into two or several smaller nuclei. These reactions are used in atomic reactors.

Let us come back to the simplest atomic reaction: the arrival of a proton or neutron in the initial nucleus. The neutron reactions are accomplished easily—there is nothing to prevent the neutron from approaching the nucleus so close that the nuclear forces, which act only at small distances, 'pull' it into the nucleus. With protons it is more difficult to obtain these reactions. In fact, all nuclei have a positive charge, the same as the proton. Therefore, as the nuclei repel the protons large energies are needed to overcome the electrostatic force of repulsion and allow the protons to approach the nucleus at a distance where the strong, but limited in time, nuclear forces can act. There are almost no free neutrons in space. It is true that sometimes in the process of nuclear reactions neutrons are formed, but this case is rare although important for the construction of heavy elements (see below).

Nuclear reactions with protons should be more frequent in cosmic conditions as hydrogen is the most widespread element in the Universe. Free protons exist everywhere, including stellar interiors. The temperature in the central parts of stars is high and therefore there are many protons with large velocities. Indeed the temperature is also an 'accelerator' of protons and alpha particles in stars. However, we must note that the temperature in stellar interiors is still not high enough to assure for each proton the possibility of entering into a nuclear reaction. As a result of numerous collisions between protons any one of them can occasionally acquire a velocity a few times greater than the mean velocity of thermal motion at a given temperature and therefore will be able to enter into the nuclear reaction, overcoming the repulsion of the positively charged nucleus. Reactions where heating to high temperatures is used to overcome the repulsion of nuclei with the same charge are called thermonuclear reactions. They were recently discovered as sources of stellar energy. One of the important technical problems of modern times is to learn how to use thermonuclear energy sources on Earth.

Having completed this short introduction we can begin to study thermonuclear reactions in stars. In such reactions, energy is released when four protons combine into one helium nucleus (in this case two protons should turn into neutrons). Such a combination of protons forming a helium nucleus can take place in different ways, but the result is always the same. Let us verify that in this way large amounts of energy can really be released. The mass of one proton in atomic units equals 1.00813. Thus the mass of the four protons will equal 4.03252. However, we know that the mass of a helium nucleus in atomic units equals 4.00389. Consequently, the excess of mass, equal to 0.02863 of an atomic unit of weight, should turn into released coupling energy. The energy excess in the generation of a helium nucleus can be calculated with the well-known Einstein formula: $E = mc^2 = 1.67 \times 10^{-24} \times 0.02863$ (3 $\times$ $10^{10})^2 = 4.3 \times 10^{-5}$ ergs for one nucleus. Remember that $c = 3 \times 10^{10}$ cm/s is the velocity of light.

We shall now calculate the energy released by thermonuclear reactions in stars. The values given above show that when four protons join into one helium nucleus about seven thousand units of the mass are released and transformed into energy (0.02863/4.03252 = 0.007). Consequently, if the whole Sun were composed of hydrogen, in its transformation into helium an amount of energy would be released which would equal $2 \times 10^{33} \times 0.007 \times 9 \times 10^{20} = 1.3 \times 10^{52}$ ergs. Since the Sun radiates 3.8×10^{33} ergs every second, the transformation of hydrogen into helium would be sufficient to maintain the solar radiation at its present level for 3×10^{18} s (about a hundred milliard years). This is more than enough. Thus we have shown that the combination of protons into helium nuclei can entirely guarantee the necessary magnitude of stellar energy sources. There now remains to study the different ways of transforming hydrogen into helium.

At present two systems of consecutive reactions are known in which four hydrogen nuclei can form a helium nucleus (evidently the possibility of a simultaneous collision of the four protons and their immediate combination into a helium nucleus is negligible). The first group of reactions is called the proton sequence (the origin of this name will be explained later on) and the second group of reactions is called the carbon–nitrogen cycle.

Let us consider the proton–proton nuclear reactions. If two protons collide, in most cases they are simply deflected in different directions. Nevertheless, in very rare cases both protons will start a nuclear reaction and form a deuterium nucleus D^2 composed of one proton and one neutron. In order to make this possible two extremely rare conditions must be simultaneously realized. First, the energy of one of the reacting protons should at least exceed the mean thermal energy of the particles of the stellar medium (this is necessary for the protons, once they have surmounted the Coulomb law repulsion, to approach a distance at which the nuclear forces could react). Second, it is essential that within the short time span of the protons at a close distance (10^{-21} s!) one of them should turn into a neutron, expelling a positron and a neutrino (since two protons cannot form a stable nucleus). The generated neutron joins with the proton and forms a deuterium nucleus.) The positron (a particle with a mass equal to the mass of an electron, but with a positive charge) escapes somewhere (not very far away from where it generated), joins a free electron, and the two of them are converted into two quanta of electromagnetic radiation. We repeat that an association of two very rare events is extremely infrequent—on the average this might happen for each proton once in ten milliard years! However, there are many protons in stellar interiors and therefore this reaction occurs with sufficient intensity.

The deuterium nucleus which has been formed rapidly enters a new reaction; after a few seconds it meets a sufficiently rapid proton with an energy allowing it to come close enough to the deuterium nucleus to join it. As a result one nucleus of helium isotope is formed with an atomic weight equal to 3, and is composed of two protons and one neutron—He^3. The great rapidity of this reaction (compared to the first one) can be explained by the fact that the proton is not transformed into a neutron, as in the first case. Moreover, deuterium is generally very active in nuclear reactions. Let us remark that modern thermonuclear reactors on Earth also use the activity of deuterium. It is evident that in this reaction an excess of energy appears which leaves the nucleus by electromagnetic radiation.

The further destiny of the nucleus of a helium isotope He^3 can be different, depending on the temperature and the presence of a helium isotope He^3 in the stellar matter, but the result is always the same, i.e. the formation of a helium nucleus. We shall first consider the most simple way. The nucleus of a helium isotope He^3 (already having three protons, one of which is transformed into a neutron) meets another nucleus of the same kind and enters into a reaction with it. As a result of this reaction the two He^3 nuclei form one nucleus of an ordinary helium isotope He^3 and two protons (from the initial six) are again released. It is true that this reaction is also quite rare—one nucleus of a helium isotope looks for a 'partner' for the reaction for about one million years. The scarcity of reactions is explained by the fact that there are much less He^3 isotope nuclei than protons and therefore the probability of an encounter with a partner possessing enough energy to surmount the repulsion is very small.

In other, scarcer, varieties of the proton–proton sequence of reactions, the nucleus of a He^3 isotope combines with an ordinary He^3 nucleus, the result being a beryllium nucleus Be^7. This nucleus in its turn can capture a proton and form a nucleus of boron (B^7) or can capture an electron and turn into a lithium nucleus. The lithium capturing a proton turns into beryllium (Be^8) which rapidly decays into two alpha particles ($2He^4$). The same thing happens with a boron nucleus but in the beginning a positron and a neutrino are released ($B^8 \rightarrow Be^8 + \beta^+ + \nu$). These 'boron neutrinos' are being searched for in 'neutrino astronomy'. The release of energy in different varieties of the proton–proton sequences is not the same and must be taken into account in precise calculations.

As a final result, the four protons formed one nucleus of an ordinary helium isotope (two protons being transformed into neutrons). In all these reactions energy has been released either in the form of neutrinos and positrons or in the form of electromagnetic radiation, or finally part of the energy was released with the protons in the last reaction. The energy of the positrons and of gamma rays and the kinetic energy of protons is rapidly transformed into thermal energy of the stellar matter and only the neutrinos can transport their energy through the whole star and escape into space. Let us write the fundamental proton–proton sequence as a formula (the origin of the same of this branch does not need to be explained):

$$(12) \quad \begin{array}{ll} H^1 + H^1 \rightarrow D^2 + \beta^+ + \nu & \text{tens of milliards of years} \\ D^2 + H^1 \rightarrow He^3 + \gamma & \text{a few seconds} \\ 2He^3 \quad\;\; \rightarrow He^4 + 2H^1 & \text{a few millions of years} \end{array}$$

Here β^+ stands for the release of a positron during the reaction process, ν represents the release of a neutrino, and γ the radiation of an electromagnetic energy quantum.

On the right-hand side we give the characteristic time spans during which one hydrogen, deuterium, and helium nucleus searches for a partner for the reaction. The rapidity of reaction depends on the density, on the percentage abundance of hydrogen and helium (this also depends on the predominant reaction sequence), and even more on the temperature. The increase of each of these parameters brings about an increase in the rapidity of the reaction—the reader can imagine the cause of these dependencies. The time spans given above correspond to conditions which exist in stellar interiors, such as the Sun or slightly cooler stars.

At the end of each reaction cycle about the same amount of energy is released—4.0×10^{-5} erg(0.3×10^{-5} erg is carried off by neutrinos)—but as the probability of reaction changes at different temperatures and densities of the stellar matter, the amount of energy (in ergs) emitted in one second by one gram matter (we shall denote this value by ε) also depends on the temperature and the density. The deduction of the formula describing this dependency is simple, but cannot be given here. In fact the probability of reaction depends on the so-called 'tunnel' effect—the surmounting of the potential barrier by the proton. This can be understood only on the basis of quantum mechanics which is beyond the scope of a school physics course. For the informed reader understanding quantum mechanics we recommend specialized literature and only give the final formula written in logarithmic form:

$$(13) \quad \log \varepsilon = 6.3 - \frac{14.8}{T_6^{1/3}} + \log \frac{\rho X^2}{T_6^{2/3}}$$

Here the temperature is given in millions of degrees (shown by the index '6', the number of zeros left out), ρ stands for the density in grams per cubic centimetre, and X is the relative abundance of hydrogen—a parameter we have already met in Chapter 1. This formula shows the rapid increase in energy release with the increase in temperature and the slower increase with the increase in density and the percentage of hydrogen abundance. The reader already knows the reason for this. The dependence on the abundance of the He^3 isotope is not evident here, but in fact it is 'hidden' in the first term.

In more precise calculations detailed tables are used instead of formula (13), whereas in approximate calculations even this formula is simplified by a relation of the following kind:

$$(14) \quad \varepsilon = \varepsilon_0 \rho X^2 T_6^n$$

where ε_0 and n are certain numbers specially chosen so that formula (14) yields the same result as (13) or the corresponding tables. For this reason ε_0 and n depend on the temperature, although only slightly. For example, at a temperature of the matter lower than ten millions of degrees the exponent in (14) is close to 5, at higher temperatures ($T_6 > 10$ million degrees) this exponent diminishes to 4, and at temperatures of a few tens of millions of degrees the exponent $n \simeq 2$. In other words, the values ε_0 and n are considered as constant only in determined and not very large temperature spans. The fact that the parameter n decreases with an increase of T_6 can be easily understood. Indeed, at relatively low temperatures there is a very small number of rapid protons capable of surmounting the Coulomb barrier of proton repulsion by the nucleus. With the increase in T_6 the number of such protons increases exponentially; at very high temperatures nearly every proton surmounts this barrier, in which case the dependence of the energy release on the temperature becomes weak.

We shall now consider whether the proton–proton reaction can assure energy radiation from the Sun. We admit that in a certain region near the centre of the Sun the following parameter values exist: $T = 14$ million degrees, $\rho = 100$ g/cm^3, $X = 0.71$. Using formula (13) we then obtain $\varepsilon = 13$ erg/(g s). With a mass of 2×10^{33} g the Sun emits 3.8×10^{33} erg/s, that is 1.9 erg/(g s). Subsequently, if a seventh of the mass of the Sun has properties close to the ones described above then the thermonuclear reactions of the proton–proton sequence can assure the actual luminosity of the Sun. If the temperature in the interior of the Sun is higher, then the region of energy generation can be considered to be of a smaller dimension. However, if, in reality, the temperature of the Sun is smaller, the situation is bad and the thermonuclear reactions cannot assure the luminosity of the Sun. We shall come back to this when considering 'neutrino astronomy'.

For stars of larger mass than our Sun the proton–proton sequence does not yield a sufficient energy release. In this case a more effective mechanism at high temperatures enters into action—the carbon–nitrogen cycle—which we will consider now.

In the proton–proton sequence only protons react, whereas in the carbon–nitrogen cycle carbon, nitrogen, and oxygen nuclei also take part in the reactions. Higher temperatures are needed in these reactions than for the proton–proton reactions since the greater charge of these nuclei demands greater energy of the protons in order to overcome the Coulomb repulsion. On the other hand, the first reaction of this cycle (at a sufficient temperature) is much more probable than the first proton–proton reaction since it is not related to the necessity of transforming a proton into a neutron. For this reason the carbon–nitrogen cycle is much more effective than the proton–proton reaction at temperatures higher than 15–16 million degrees.

We give here the formulae of successive reactions of the carbon–nitrogen cycle:

$$(15) \quad \begin{array}{ll} 1. \ C^{12} + H^1 \rightarrow N^{13} + \gamma & \text{tens of millions of years} \\ 2. \ N^{13} \qquad\ \rightarrow C^{13} + \beta^+ + \nu & \text{7 minutes} \\ 3. \ C^{13} + H^1 \rightarrow N^{14} + \gamma & \text{a few millions of years} \\ 4. \ N^{14} + H^1 \rightarrow O^{15} + \gamma & \text{hundreds of millions of years} \\ 5. \ O^{15} \qquad\ \rightarrow N^{15} + \gamma + \nu & \text{82 seconds} \\ 6. \ N^{15} + H^1 \rightarrow C^{12} + He^4 & \text{hundreds of thousands of years} \end{array}$$

Also given is the characteristic time during which the fundamental reacting nucleus 'looks for' a proton with a velocity sufficient for a thermo-nuclear reaction (reactions 1, 3, 4, and 6) to take place or a time span during which the unstable isotope which is formed decays and releases a positron and a neutrino (reactions 2 and 5). The cycle starts when a rapid proton is captured by a carbon nucleus and is kept there by nuclear forces. An unstable nitrogen isotope is formed with an atomic weight equal to 13 (too light for a normal nitrogen nucleus) which disintegrates within 7 minutes and turns into a stable carbon isotope with the same atomic weight. This nucleus twice captures successively a rapid proton (each time releasing the energy excess by way of electromagnetic radiation) and is transformed into an unstable oxygen isotope with an atomic weight equal to 15 (too light for an ordinary oxygen nucleus). After only 82 s it disintegrates, releasing as usual a positron and a neutrino, and forms a stable nitrogen isotope. Finally, having captured one more rapid proton, the fourth in the reaction sequence, this isotope gives a helium nucleus and the initial reactant, an ordinary carbon isotope. The cycle then starts again from the beginning. In this cycle the carbon and nitrogen are only catalysts, the final product of the reaction being the same as at the start; the four protons formed a helium nucleus and the corresponding energy release took place (giving 4.0×10^{-5} ergs for one cycle).

All the reactions of this cycle, in which protons are captured, are slow—a few millions of years. This can be explained, as already noted, by the fact that the charges of the carbon and nitrogen nuclei are large and therefore the probability of encountering a sufficiently rapid proton to overcome the repulsive forces is small. We must also note that here the velocities of the reactions depend not only on the hydrogen abundance but also on the carbon and nitrogen abundance, which we designate by X_{CN}. As we know, in the stellar matter the mean carbon and nitrogen content is about two hundred times smaller than that of hydrogen: $X_{CN} = 0.005X$.

A variety of the cycle also exists where the nitrogen nucleus (N^{15}) capturing a proton does not disintegrate into nitrogen and carbon, as described above, but stays in the form of an oxygen nucleus (O^{16}) which thereupon undergoes the following transformations:

$$O^{16} + H^1 \rightarrow F^{17} + \gamma$$
$$F^{17} \rightarrow O^{17} + \beta^+ + \nu$$
$$O^{17} + H^1 \rightarrow N^{14} + He^4$$

In this case, the four protons form one helium nucleus while oxygen, nitrogen, as well as fluorine only serve as catalysts. However, this variety of the cycle has no practical meaning because for only one in two thousand proton captures does the oxygen atom O^{16} have a change of being preserved.

The formula to calculate the energy release in a carbon–nitrogen cycle has the same form as formula (13), but of course with different values of the constants:

$$(16) \quad \log \varepsilon = 28.2 - \frac{66.8}{T_6^{1/3}} + \log \frac{\rho X X_{CN}}{T_6^{2/3}}$$

Here again the temperature is expressed in millions of degrees. This formula shows that in the carbon–nitrogen cycle the temperature dependence of the energy release is much stronger than in the sequence of proton reactions. This is not surprising. The large positive charge of the carbon and nitrogen nuclei strongly raises their 'Coulomb barrier' and the protons need great energy to overcome it.

As before, formula (16) can be replaced by the approximate relationship (14). Now the exponents n will be even greater. With T_6 equal to some tens of millions of degrees we have $n \approx 23$; at a temperature of up to thirty million degrees this exponent decreases to 16. We now use formula (16) in the carbon-nitrogen cycle to calculate the energy release in the Sun. For the same conditions as in the case of proton reactions (and with $X_{CN} = 0.003$) we obtain the energy release $\varepsilon = 0.1$ erg/(g s). This of course is not enough—the temperature in the centre of the Sun is too small for carbon–nitrogen cycle reactions to take place. We then determine the energy output in the hot and bright star. We assume $T_6 = 30$ millions of degrees, $\rho = 10$ g/cm³, $X = 0.7$, $X_{CN} = 0.003$. We obtain $\varepsilon = 1600$ erg/(g s), whereas on the average this star emits 128 erg/(g s) for 1 gm of matter. Here, less than one-tenth of the star is sufficient to assure its luminosity in the given conditions.

The strong dependency of energy release in the carbon–nitrogen cycle on temperature exhibits the great need for precision in its determination—a precision not given by formula (5). It is better to do the opposite, i.e. knowing the luminosity of the star and supposing that energy is released, e.g. in 1/200 of its mass, we can use the given formulae to determine the central temperature of the star. We leave the reader to do these calculations and check that the central temperatures are close to those given in Table 3.

Thus we can see that the thermonuclear energy sources assure the luminosity of all stars of the main sequence. Of course we are not yet convinced of the trustworthiness of all the data as most of the reactions were observed in laboratories. Besides, since in laboratories more rapid protons are used for these reactions than those reacting in stellar conditions (we cannot wait for millions of years), the experimental data must be extrapolated for small particle energies and in this case serious errors are possible. It is possible, for example, that for certain determined proton energies the reaction is particularly intense (in this case there is a 'resonance'). Some of the reactions are such that, for example, the first of the proton–proton reactions cannot be realized in laboratory conditions. It is therefore possible that later on the numerical value of the energy release could change, but the variation will not be much larger. As for the thermonuclear reactions, their sequences and their role in the process of energy production in the main sequence stars are beyond any doubt.

Do these reaction sequences exhaust the list of thermonuclear reactions which can serve as sources of stellar energy? Generally speaking, no. First of all, at stellar temperatures thermonuclear reactions with light elements such as lithium, beryllium, and boron are very intense. In the end, hydrogen, together with the nucleic of these elements, is transformed into helium. However, in these reactions the lithium, beryllium, and boron nuclei are not renewed as was the case with the carbon nuclei in the carbon–nitrogen cycle. In other words, if in the carbon cycle the carbon were the catalyst, and therefore would not be totally consumed in the entire process, then in a reaction with light elements these elements are 'burned' and turn into helium. For this reason these elements are rapidly (within about a million years 'burnt' in stellar interiors. They serve as a source of stellar energy only during a short (of course relatively) initial period in the stellar evolution.

At temperatures of some hundreds of millions of degrees, when hydrogen is already entirely 'burned', the source of stellar energy can be the so-called triple alpha process. This process can be resumed as follows. If two alpha particles with large energies (ten times greater than the thermal energy at a hundred million degrees) collide, they can, for a very short time, form an unstable nucleus of a beryllium isotope with an atomic weight equal to 8. If, after a time, before this nucleus manages to decay inversely into alpha particles, one more alpha particle joins it, then a stable nucleus of the ordinary carbon isotope C^{17} can be formed and a great amount of energy is released. In this reaction the 'fuel' is helium and the product of 'combustion' is carbon.

The formula for energy release in the triple alpha process differs a little from the corresponding formulae for the proton reactions:

$$(17) \quad \log \varepsilon = 17.5 - \frac{1890}{T_6} + \log \frac{\rho^2 Y^2}{T_6^3}$$

where Y represents the helium abundance. The difference in temperature dependence can be explained by the fact that for the alpha particles in the reaction we must entirely determine the energy corresponding to the already mentioned resonances.

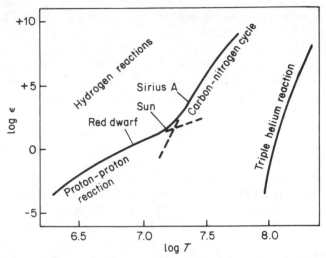

Figure 6 Thermonuclear energy release in different reactions with the following assumptions: the amount of hydrogen in hydrogen cycles is 100 g/cm^3, the amount of carbon and nitrogen is two hundred times smaller in weight, and the amount of helium is 10 000 g/cm^3 (in the triple helium reaction)

The triple alpha process can only give an important energy release at temperatures exceeding 100–120 millions of degrees. It is interesting to note that if we try to replace formula (17) by formula (14) then at $T_6 = 100$ million degrees we obtain $\varepsilon \sim T^{41}$! In ordinary stars of the main sequence the temperature does not reach a hundred degrees and therefore this reaction is not important for them. However, at certain stages of stellar evolution, as we shall see later, the triple alpha process can be a fundamental source of stellar energy.

Figure 6 shows a graph (on a logarithmic scale) with the dependence of the energy release on temperature for all three sequences of reactions examined. We assumed that in the hydrogen reactions $\rho X^2 = 100$ g/cm^3 and in the helium reaction $\rho^2 X^3 = 10^8$ g^2/cm^6. We have shown the relative role of the reactions in stars which are of different types but all belong to the main sequence. This graph clearly illustrates the quite strong dependence of the energy release in a thermonuclear reaction on the temperature. At higher temperatures reactions with heavier elements start. It is true that at the moment the maximum temperature in stellar interiors is now known but we presume that there are temperatures of up to ten milliard degrees. What happens then?

Already at $T_6 = 100$ million degrees an important reaction starts:

$$(18) \quad C^{13} + He^4 \rightarrow O^{16} + n$$

where n represents a neutron Its significance is not so much the fact that energy is released in this case but that the neutron which appears in the reaction can 'stick' to any other nucleus and in this way increase its atomic weight—all heavier elements can be formed successively in this way. There are also other reactions in which neutrons are released, for example $C^{12} + C^{12} \rightarrow Mg^{23} + n$.

Of course in this case a much higher temperature is needed (up to one milliard degrees).

The successive formation of heavy elements by addition of neutrons can be realized in different ways. Theory has predicted the existence of so-called s-processes (slow) and r-processes (rapid). In the s-processes the following happens. The atomic nucleus captures a neutron and is unstable with respect to the beta-decay, which is usually slow. If the time span between the successive captures of neutrons is also small (small quantity of neutrons) the nucleus can undergo a beta-collapse, release an electron, and so increase the atomic number of the nucleus forming a new stable atomic nucleus. Such a successive formation of elements can take place in hot and dense stationary stars; it is in this way that the elements in the middle of Mendeleev's table are probably formed. On the other hand, if there are many neutrons and the time span between successive captures of neutrons is smaller than the period of a beta-decay, then their is a formation of heavy elements by the r-process. In this process heavy elements are formed, shown at the end of Mendeleev's table.

There are also p-processes in which the heavy atomic nuclei successively capture free protons. Of course, to make this process effective there should be, on the one hand, a sufficiently high temperature (2×10^9 degrees) and, on the other hand, free protons which burn at much lower temperatures.

Usually it is assumed that the r-processes and p-processes can take place only in explosions of stars—the phenomenon of supernovae which we shall study in Chapters 8 and 10. In such bursts the matter of the central parts with high temperatures and hydrogen already burnt up can mix with the matter from exterior stellar layers where the temperatures are much lower.

In stationary stars heavy elements can also be formed by the successive addition of helium nuclei: $C^{12} + He^4 \rightarrow O^{16} + \gamma$; $N^{14} + He^4 \rightarrow F^{18} + \gamma$; $O^{16} + He^4 \rightarrow Ne^{20} + \gamma$; $Ne^{20} + He^4 \rightarrow Mg^{24} + \gamma$; etc. Ne^{20} and Mg^{24} are formed only in stars with masses bigger than $30M_\odot$. The abundance of O^{16} can reach 50 per cent., the abundance of neon being not more than 20 per cent.

If very high energies occur in stellar interiors, energy release will also take place in reactions between heavy elements:

$$(19) \quad \left.\begin{array}{l} C^{12} + C^{12} \rightarrow Na^{23} + H^1 \\[4pt] C^{12} + C^{12} \rightarrow Ne^{20} + He^4 \end{array}\right\} T > 8 \times 10^8 \text{ degrees}$$

$$\left.\begin{array}{l} O^{16} + O^{16} \sim S^{32} + \gamma \\[6pt] 2Ne^{20} \rightarrow O^{16} + Mg^{24} \end{array}\right\} T > 1.3 \times 10^9 \text{ degrees}$$

However, all thermonuclear reactions with an energy release end in the formation of iron nuclei Fe^{56}. In order to form a more heavy element with this non nucleus it is necessary to lose more energy than is released in the reaction process. For this reason heavier elements are formed only by the s-process.

Consequently they are rare. Meanwhile it is possible that stellar bodies exist which are composed almost exclusively of oxygen, neon, magnesium, and even iron.

If we continue to increase the temperature of such 'stars' then the iron nuclei will decay into alpha particles. This will happen at a temperature higher than 7–8 milliard degrees. In this way we approach the second important consequence of the activity of thermonuclear reactions—the change in chemical composition of the star—and at the same time deal with the question of the origin of chemical elements.

This is a very complex problem which goes beyond the limits of the stellar theory treated here. What was the initial composition of the matter that formed the stars? Because there are few heavy elements in old stars (see Chapter 1) we can presume that their abundance was very small or even absent in the original matter. Heavy elements are quite abundant in young stars and in interstellar space so they must have formed in stars which had to go through a high temperature phase.

The question of helium abundance in the original matter is very interesting. The modern forms of the 'hot universe' model need an initial abundance of helium representing 30 per cent of the weight. As helium is also the first product of thermonuclear reactions there should be a large amount of helium. Unluckily helium is a very difficult element to observe and at present very little data are available to judge its real abundance.

We have already remarked that light elements such as lithium, beryllium, and boron burn quite rapidly (within millions of years) in stellar interiors and turn into helium isotopes He^3 and He^4. In thermonuclear reactions these elements are not renewed. Observations have shown that in stars with an important intermixing there is very little lithium. Of course, there are stars with large abundances of lithium, but this means that there is no intermixing of matter and lithium has been conserved in external layers with a low temperature.

On the other hand, it is known that lithium, beryllium, and boron nuclei are abundant among cosmic ray particles. Effectively, the nuclear reactions with rapid heavy particles of cosmic rays can bring about the formation of a sufficient amount of these nuclei. It is now considered that lithium, beryllium, and boron, actually existent, originated only in cosmic rays. They diffused into the interstellar medium and from there passed into stars.

The elements carbon, nitrogen, and oxygen take part in the carbon–nitrogen cycle. With the passage of time, a relative number of nuclei of these elements will change until an equilibrium state is reached, that is C : N : O are inversely proportional to the probability of reactions in which these elements take part. In fact, the greater the probability of these reactions, the more corresponding nuclei turn into other elements. It appears that if the temperature in the centre of the star exceeds 16 million degrees, almost all carbon and nitrogen atoms turn into N^{14} isotopes (an abundance of 95 per cent.) The C^{12} isotope represents 4 per cent and the C^{13} isotope 1 per cent. However, the general amount of carbon, nitrogen, and oxygen nuclei does not change in the carbon–nitrogen cycle process. For this reason the carbon–nitrogen cycle also changes the total content of hydrogen and helium.

The fact that the relative abundance of carbon, nitrogen, and oxygen should be inversely proportional to the probability of the carbon–nitrogen cycle reaction can be checked by observations. In many stars this condition is in fact realized, but there are quite numerous exceptions. There are stars with many different carbon isotopes (this interesting phenomenon was discovered in 1948 by the Soviet astronomer G. A. Shain) and we know even more striking anomalies of the chemical composition of stars—the stars of class Ap—which have not yet been explained.

Certain reactions have been shown in general and processes mentioned where the formation of heavier elements could occur. In recent times a large number of publications have appeared in astrophysics and physics where these possibilities have been analysed. However, it must be said that it is still difficult to trace with certainty the entire evolution or formation of all elements, but it is possible.

Undoubtedly, carbon, nitrogen, and oxygen are formed in the helium burning stage in hot nuclei of stars which are not too massive ($M > 0.5\ M_\odot$). In the interiors of more massive stationary stars other heavy elements, apparently up to neon, magnesium, and possibly flint, develop through successive captures of helium nuclei.

It is not very clear whether a sufficient amount of iron can be obtained in stationary stars. If at any stage in stellar evolution the temperature in the stellar interior reaches 3–5 milliard degrees for a density of matter of 10^5–10^8 g/cm^3, the iron nuclei and close elements (nickel, cobalt) are formed in a sufficient amount. However, it is more probable that these elements are formed in stellar outbursts.

Heavier elements with atomic weights exceeding 60 can be formed only in the processes of neutrino capture (s- and r-processes) and proton capture (p-process). This probably happens in supernovae flares, when in stellar interiors regions appear, though very briefly, with very high temperatures up to 4–5 milliard degrees and very large densities up to 10^6–10^8 g/cm^3. Theoretical calculations have shown that the observed abundance of chemical elements formed in supernovae flares can be explained if the temperature and density of the matter are correctly chosen. One more possibility of heavy element formation is the accretion of hydrogen-rich matter on the surface of a very dense star—a white dwarf with no hydrogen. Such formation exists through proton capture (p-process).

Thus, we can see that the formation of chemical elements occurs principally in stellar flares and other processes involving outbursts. It has not been possible to study stellar outbursts fully yet so we really know very little of the origin of the chemical composition of the universe.

Let us pass on to the very popular 'neutrino astronomy'. As we already know, most of the energy released in thermonuclear reactions is carried away by protons—quanta of electromagnetic waves. An important part of the energy leaves with positrons or is simply transmitted by nuclei in the form of kinetic energy. Most of this energy is immediately converted into another form, but, as in the proton sequence and the carbon–nitrogen cycle, about 10 per cent. of the energy release passed to the neutrinos. This energy cannot be converted into thermal energy. The neutrinos have a great capacity of penetration and though they are formed in the centre of the star they freely pass through its whole thickness, carrying off to the exterior the energy they have received. For this reason we can speak of the 'neutrino luminosity' of a star.

Therefore the thermonuclear energy leaves the star in two forms: the energy converted into heat gradually filters through the thickness of the star and comes out in the form of 'photon radiation', while the energy taken away by neutrinos leaves in the form of 'neutrino luminosity'. Neutrino luminosity is very difficult to observe as the neutrinos pass directly through the Earth without any interaction and so do not register on instruments. Although difficult, it is not impossible, and the problem of neutrino luminosity must be considered.

Since the neutrino luminosity of ordinary stars does not exceed a few per cent. of their optical luminosity, one can only hope to measure the flux of neutrinos coming from the Sun. The total amount of neutrinos formed can easily be estimated: the formation of one helium nucleus releases about 4×10^{-5} erg of energy and two neutrinos appear. In one second the Sun emits almost 4×10^{33} erg and consequently in one second 10^{38} helium nuclei and 2×10^{38} neutrinos are formed—almost all of these leave the Sun. This means that on each square centimetre of the Earth's surface 8×10^{10} neutrinos fall per second, which is such an enormous amount that it is difficult for us to grasp.

The measures of the whole neutrino flux do not give us very much information and perhaps only upset all notions of thermonuclear reactions if these neutrinos do not appear. There is one particularity. We have already mentioned that in stellar interiors certain varieties of thermonuclear cycles occur and in different reactions neutrinos with different energies are released. In particular the above-mentioned reaction of the transformation of boron into beryllium gives a high energy capable of turning a chlorine isotope Cl^{37} nucleus into the nucleus of a radioactive argon isotope Ar^{37}. This reaction can only be due to boron neutrinos as only they have enough energy.

The amount of boron neutrinos released by the Sun is some ten thousand times smaller than that of ordinary neutrinos, but the reaction of B^8 formation depends strongly on temperature and on helium abundance. For this reason, the result of boron neutrino flux measurements permits us to determine the helium abundance in the interior of the Sun and also its central temperature. In 1968 an attempt was made to detect boron neutrinos. The result was important, showing that the boron neutrino flux falling on the Earth's surface is much less than 2×10^6 particles per square centimetre in one second. Since then measurements of boron neutrino flux have regularly been made (though only in one laboratory, by the American physicist Davison). As one measurement (the accumulation of the radioactive Ar^{37} isotope) lasts for about a hundred days, during the whole period up to 1976, only about forty measurements have been carried out. In most cases it was not possible to confirm the existence of the boron neutrinos. An estimation of the experimental precision shows that in this case the upper limit of the neutrino flux from the solar interior is six times smaller than the value given above. In 1975–1976 Davison finally managed to discover these particles (with a magnitude of the flux two times smaller than the theoretical one) but at the end of 1976 the next measure again yielded a negative result.

There are a few ways of explaining this negative result, but they are all linked with a serious revision of some fundamental ideas in the theory of inner stellar structure. The absence of boron neutrinos means that the probability that the reaction $He^3 + He^4 \rightarrow Be^7$ occurs is much smaller. This can be the case, for example, if the temperature of the matter in the solar interior is smaller than 14.4 million degrees, as shown by calculations. We could decrease the numerical value of the central temperature, supposing that there is a rapidly rotating nucleus. Then the centrifugal force would permit us to slightly decrease the gas pressure. It is difficult to conceive why this rotation does not slow down. We could suppose that the temperature in the solar interior varies periodically with time; in that case the values given above only determine the mean temperature and actually the temperature in the centre of the Sun could be a little lower. The cause of these fluctuations is not clear. We could admit that in the centre the helium abundance is simply small ($Y < 0.22$), but the age of the Sun is great, about 5×10^9 years, and so a large amount of helium should have accumulated. Of course we could assume that in the solar interior, notwithstanding the absence of convection, strong intermixing occurs which impedes the accumulation of helium. On the other hand, the whole modern theory of stellar evolution corresponds better to the hypothesis that there is no important intermixing in stellar interiors. We could also assume that the abundance of heavy elements in the centre of the Sun is much smaller than at its surface (that is $Z < 0.02$), but it is difficult to explain how such a composition could arise.

An attempt to explain the negative result of the neutrino experiment by the revision of the neutrino theory has been made; e.g. it has been assumed that a neutrino can decay on its way from the Sun to the Earth (it traverses this distance within 8 minutes). A solution to this problem has not yet been found, but it is hoped that this difficulty will be overcome without a wholesale reconsideration of the theory of the inner structure of the Sun and subsequently of stars in general.

Neutrinos with great energy are formed in the interior of the Sun in the rare reaction $p + p + e^- \rightarrow D^2 + \nu$. These neutrinos are also captured by the nuclei of chlorine isotopes Cl^{37}. Their abundance is about a hundred times less than that of boron neutrinos, but since measurements can now be made with a precision ten times greater than previously it should be possible to detect the presence of these neutrinos. However, even if they do not appear, there will have to be alterations made to the whole theory of inner stellar structure.

In neutrino astronomy another interesting problem exists. The fact that neutrinos formed in hydrogen reactions carry off about 10 per cent. of the total energy released has little influence on the stellar structure. In the burning of helium (with the generation of carbon and oxygen) neutrinos in general do not appear. However, in the evolution of stars it is possible, at least in principle, that there are states where the neutrino luminosity can exceed the optical luminosity. Where does this lead us?

G. A. Gamov supposed that at high temperatures in stellar interiors an important part can be played by phenomena which he called the URCA-process. Gamov explains this denomination in the following way. In Rio de Janeiro there is a casino URCA where the players imperceptibly lose their money at roulette. The same thing happens with stars (which also turn) which imperceptibly lose their energy by cyclic processes of neutrino emission. In these processes an important part of the energy is carried from the star by neutrinos without having any effect on the outer layers of the star. Let us give a few examples of such a URCA-process.

The nucleus of a helium isotope (He^3) absorbs an electron with an energy of 3×10^{-4} erg (such electrons are abundant at a temperature of two hundred million degrees) and, releasing an antineutrino, turns into a hydrogen isotope—tritium (T^3). This isotope then turns into He^3, releasing an electron and a neutrino. Everything returns to the initial state, but a neutrino–antineutrino couple carries off an energy of about one hundred million of the share of an erg. Of course at two hundred million degrees, probably no He^3 nuclei are left, but in the process described energy is carried away through the whole thickness of the star.

The second example of a URCA-process has similar transformations of N^{14} into C^{14}. For this to take place there must be an electron energy of about 2.4×10^{-7} erg (a temperature of about two milliard degrees). The inverse transformation of C^{14} into N^{14} is very slow—the half-life is 5600 years.

In the case of rapid contraction URCA-processes are possible with elementary particles since neutrinos are also formed during the collapse. For example, a neutron decays into a proton, an antineutrino, and an electron. In its turn the electron is captured by a proton and forms a neutron and a neutrino. As a result the neutron remains a neutron and the neutrino–antineutrino couple carries off the energy.

Neutrinos and antineutrinos can also appear in other physical processes which are possible even if not in ordinary stars, at least in those stages of stellar evolution where the temperature increases so strongly that a large number of electron–positron couples are formed in the gas. In the annihilation of these couples many gamma quanta are formed of such great energy that in collisions with other particles their energy is divided between other newly formed particles, some of which are neutrinos and antineutrinos. It is difficult to explain the physics of these phenomena visually, but if we use a rough analogy we can say that in collisions of rapid particles (gamma quanta also being particles) very diverse 'bits' fly in all directions. Then the gamma quanta,

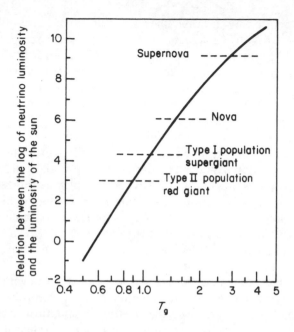

Figure 7 Growth of neutrino luminosity with the increase of central temperature. The dashed lines indicate neutrino luminosity at determined stages of stellar evolution (the lines corresponding to novae and supernovae are hypothetical)

electrons, positrons, and other particles rapidly yield their energy to the medium, i.e. they return it, in fact, inversely. The neutrino and antineutrino carry off the energy beyond the limits of the star and thus cool it rapidly. This cooling process increases as the temperature increases: at a temperature of one milliard degrees 1 g of matter loses about 10^{14} erg in one second and at a temperature of 2.5 milliard degrees the losses are already about 10^{19} erg/(g s) (at $\rho \approx 1$ g/s^3). These are large losses, greatly exceeding the energy outcome of thermonuclear reactions.

It is true that such great temperatures are probably very rare in stellar interiors. However, it appears that in a dense plasma which is not too hot neutrinos and antineutrinos can be formed when one gamma quantum disintegrates into the pair. The maximum energy release that can be obtained in this process at $\rho = 10^8$ g/cm^3 and $T = 4 \times 10^8$ degrees is approximately 10^6 erg/(g s). Here the dependence on temperature is not very strong; at $\rho = 10^8$ g/cm^3 and $T = 2.5 \times 10^9$ degrees the energy release is about 10^9 erg/(g s).

For a rough estimation of the energy release in neutrinos one can use the approximate formula

$$(20) \quad \varepsilon \approx 10^{14} \frac{T_9^8}{\rho} \quad \text{erg/(gs)}$$

where T_9 is the temperature in milliards of degrees. This formula is valid for $\rho < 10^6$ g/cm^3. At $\rho > 10^7$ g/cm^3 it is more convenient to use another formula:

(21) $\quad \varepsilon \approx 10^8 T_9^8 \quad$ erg/(g s).

Thus we would expect that if the evolution of the star is accompanied by an increase in its central temperature and density its central luminosity will increase strongly with time. Figure 7 shows an example of an increase in neutrino luminosity with an increase of the central temperature in a star whose mass is close to the mass of the Sun. The dashed lines show the neutrino luminosity at determined, though hypothetical, stellar evolution stages. In the case of supernovae the neutrino luminosity can exceed the optical luminosity of the Sun by many milliards of times. We have no way of checking this result by observation. Moreover, the duration of neutrino luminosity of supernovae lasts for only about 20 ms.

The possibilities of discovery of neutrinos were studied by the Soviet physicist B. Pontecorvo, who can be considered as one of the founders of neutrino astronomy.

5

Dense stars—white dwarfs

In the description of different stellar sequences we have seen a particular group of stars, called white dwarfs, characterized by their great densities. Due to this the stellar matter in white dwarfs has particular characteristics. If we try to calculate the temperature in the centre of a white dwarf using formula (5), we obtain hundreds of millions of degrees, since the masses of these stars are comparable to the masses of ordinary stars and the radii are ten times smaller. At such temperatures a large amount of thermonuclear energy should be released and it is found that the luminosity of white dwarfs is small—much smaller than the luminosity of ordinary stars of the same mass.

It is evident that the temperature in the innermost part of a white dwarf cannot be determined using formula (5). Which one of the two physical laws on which this formula is based cannot be applied to the stellar matter of white dwarfs? We cannot doubt the applicability of the universal gravitational law in general. We must therefore assume that the stellar matter of white dwarfs does not obey Clapeyron's law of gas composition. On the other hand, it is evident that it cannot be solid (in the ordinary meaning of this word) or fluid, since at the surface of white dwarfs the temperature already reaches 10 000 degrees and in the interior it would, of course, be greater. Moreover, the density of solid and fluid matter, where atoms are densely 'stored' and accompanied by their electron envelopes, does not exceed 20 g/cm^3. As the density of white dwarfs is ten thousand times greater, the distances between the centres of atoms in the stellar matter of white dwarfs should be several tens of times smaller than in ordinary solid and fluid objects in which—we emphasize this—atoms are 'contiguous'. For this reason, in the stellar matter of white dwarfs the atoms would be destroyed and the electrons separated from the nuclei. Such matter cannot be said to be in a solid or fluid aggregate state, because a solid object and a fluid can only exist when the atoms are entirely conserved (the particularities of solid bodies and fluids are determined by the chemical particularities of entire atoms).

We call a gas a state of matter where, on the one hand, its particles are at large distances from each other (much greater than their dimension) and, on the other hand, the particles of the gas move freely in space, except for a short instant when they collide. In the matter of white dwarfs the first condition is always realized, the second only partially. In dense white dwarfs the bare atom nuclei cannot move freely; they oscillate around the equilibrium position in their 'cells' which is a peculiarity of solid bodies. However, the basic particularities of the white dwarfs' matter are defined by electrons and this 'solidness' of the nuclear component of the matter is not of great importance.

Therefore, the matter of white dwarfs is composed of different electrons and bare atomic nuclei (i.e. of the same particles as the matter of ordinary stars) and is a gas, but thanks to the high density and pressure this gas has unusual properties. Such a gas is usually called degenerate and we shall now study its properties. For this, we shall start in an altogether different way and consider an ordinary atom of a certain chemical element. We shall choose oxygen, but an atom of any element could be used. An oxygen atom is composed of a nucleus and eight electrons which move around the nucleus in different orbits. Three orbits have two electrons and two orbits one electron. To be more precise, we can say that in each couple of electrons in one orbit, one electron has one direction of spin, i.e. the direction of the proper mechanical moment, and the other the opposite. These orbits are situated at different distances from the nucleus. In the innermost orbit there are two electrons; the other electrons are situated in higher orbits. At once the question arises: why are there not more than two electrons in each orbit? Why do the electrons not jump from the higher orbits to the lower ones, releasing their stock of potential energy? We know that the electrons in atoms can freely pass from one orbit to another, emitting or absorbing energy. It appears that such transitions are only possible if in the orbit in which the electron falls there are either no electrons or only one electron. On the other hand, we know that if in a mechanical system there is a supply of potential energy (as, for example, in a lifted stone) and if the potential energy has the possibility of becoming kinetic energy (in the case of a lifted stone the possibility is to fall), it will slow down this action. This means that in the atom something impedes the upper electron from falling into the lower orbits occupied by two electrons. Why is there only room for two electrons in one orbit? This is difficult to explain clearly if one does not want to go beyond the limits of a school physics course. It is one of the laws of quantum mechanics called the Pauli principle.

We must note that this property of electrons is observed not only in single atoms but also in metals. It is known that so-called free electrons exist in metals. Each atom in a metal loses one or two outer electrons, which are the less strongly bound and can freely move in the metal. These free electrons transport the electric flux and heat, which is why metals are such good conductors. In non-conductors all electrons stay fixed in their atoms and there are no free electrons; therefore non-conductors do not conduct an electric flux. The number of free electrons in metals is either equal to the number of atoms in the metal or is twice as large. Each free electron in the metal moves on its trajectory—a certain line—with a certain velocity. On any one trajectory there cannot be more than two electrons moving with the same velocity. The spin of these electrons will also be different. If three or more electrons move on the same trajectory, they must have different energies and velocities. Of course we can ask: why must they in fact move on the same trajectory? There is a lot of space in a metal, why can they not move on other trajectories? In fact this is not possible. It appears that electrons in an atom can only move in entirely determined orbits. This is also true for free electrons in a metal; they move on entirely determined trajectories separated from one another by a certain distance. Thus there are also trajectories which are successively occupied by pairs of electrons. The following electrons must 'sit' on more 'distant' trajectories or move with greater velocities.

We must note that, rigorously speaking, there is no concept in quantum mechanics of 'determined trajectories'. One can only speak of the probability of the presence and motion of electrons in a given position in space. For this reason the reader who wants a precise expression should always understand by 'determined trajectory' the term in quantum mechanics 'determined quantum state'.

The different physical experiments and their theoretical interpretations lead to the conclusion that this new rule is a universal physical law and therefore should also be valid for stars. We now use the Pauli principle to explain the paradoxical particularities of white dwarfs.

In the beginning of this chapter we saw that due to the high density of matter in the innermost part of white dwarfs the atoms are shattered. Consequently, the electrons are free and move not in orbits around the nuclei, as in atoms, but in open, complex, and often irregular trajectories, as they do in metals. If bare nuclei can move freely they can form a gas of atomic residues. However, the motion of atomic nuclei has little influence on the properties of white dwarfs. Free electrons in a star are submitted to the Pauli principle in the same way as free electrons in a metal. Let us see what results from this. The Pauli principle changes the behaviour of matter only if the number of electrons is larger than the number of free trajectories. If there are less electrons, the Pauli principle, although still valid, has no effect—each electron can choose a free orbit and move on it with an arbitrary velocity. This is the case in ordinary stars, where there are many free electrons torn off the nuclei but even more free trajectories. The situation is different in white dwarfs. Since the densities are much higher there (and consequently in one cubic centimetre there are many more electrons than in the stellar matter of an ordinary star) not enough free trajectories are avilable for all the electrons. Consequently, the electrons in the matter of white dwarfs must occupy the same trajectories and according to the Pauli principle must move on them with different velocities.

Let us examine this stellar matter from a slightly different point of view. Let us suppose that we heated the matter of an ordinary star with a comparatively small density to millions of degrees and then contracted and cooled it simultaneously (it is necessary to cool it, for as we know contraction heats the gas). During the contraction the number of free trajectories would decrease.

On the other hand, by cooling the stellar matter, energy is taken away from the electrons and atomic nuclei and consequently their velocity would decrease. How long can the process of contraction and cooling of the gas last and what will result? We assume that we reach a state where the number of trajectories decreases to the number of electrons; from this moment all trajectories are occupied. After further contraction several electrons will move on one trajectory. According to the Pauli principle they should have different velocities. We now stop the contraction and continue the cooling. In this case, as we know, the velocity of all particles decreases, as does also the number of electrons. If we could cool the stellar matter to absolute zero ($-273\,°C$) all particles should come to a stop—as a matter of fact, the absolute zero in temperature is characterized by the fact that at that temperature all motion of atoms, molecules, and other particles ceases. But the electrons in our stellar matter cannot stop. According to the Pauli principle, as they are situated on the same trajectories they must move with different velocities. On each of the trajectories there can only be two electrons. Thus, having cooled the stellar matter we have not stopped the electrons. Let us go back again to the contracted and heated stellar matter and continue the contraction so that on each trajectory there will be many electrons; the more electrons on the same trajectory the greater will be the velocity interval of their motion. Thus, according to the Pauli principle, the velocities of the particles will become greater than their thermal velocities and the heating or cooling of this matter will not have much influence on the velocity of the electrons.

A gas in which electrons occupy all trajectories and move with great velocities (according to the Pauli principle) is called a 'degenerate electron gas'. A degenerate electron gas can have an arbitrary temperature and will stay degenerate if the velocities of the electrons are only related to the necessity of occupying the same trajectories and remain larger in comparison to the thermal velocities.

We shall now consider the question of degenerate gas pressure. Note that the pressure of ordinary gases is small at low temperatures—as a matter of fact, according to Clapeyron's formula, the pressure is proportional to the temperature. Therefore, if we could obtain a gas at a temperature of absolute zero (physically this is not possible, since at low temperatures any gas will contract) its pressure would also equal zero. It is known that pressure is an impulse transmitted to the containing limit in a collision with a gas molecule. In fact, the gas molecule (considering an ordinary gas) hitting the containing wall transmits the impulse and moves away, again transmitting the same impulse. It is evident that the greater the velocity of motion of the molecule (the greater the temperature), the greater the transmitted impulse and the greater the pressure. Stationary molecules do not transmit any impulse.

A degenerate electron gas should exert pressure on the container since the electrons move with large velocities and nothing impedes their collision with the container and the transmission of their impulse—as a matter of fact, after their repulsion from the container they continue to move with the same velocity but only in the opposite direction. Moreover, the pressure of a degenerate gas should be very large, due to the high velocities of the electrons, and since the velocities of the electrons also remain high at absolute zero temperature the pressure of the degenerate gas also remains large in this case. In general, since in a degenerate gas the velocities of the particles are not strongly related to the temperature of the gas, its pressure does not really depend on the temperature. As this is sufficient to understand the theory of white dwarfs, we shall not in general consider the influence of temperature on the pressure of an electron gas.

Of course in the stellar matter there are not only electrons but also bare nuclei—atomic residues. Note that the heaviest nuclei conserve their closest, strongly bound electrons even in the unusual conditions of the stellar matter of white dwarfs. It is evident that the nuclei and atomic residues must also take part in thermal motion and pressure. And this is the case. However the nuclei do not become degenerate, even at the highest densities which are encountered in white dwarfs, and therefore their velocities correspond to the thermal energies of the particles (much smaller than the pressure of the degenerate electron gas). For this reason the equilibrium of a star does not depend on the gas temperature of the atomic residues, but it is essential for the calculation of the thermonuclear energy release in white dwarfs.

In reference to this, we must make a remark concerning the molecular weight of the stellar matter with degenerate electron gas. As we already know, the molecular weight of the stellar matter depends mainly on the hydrogen and helium abundance. There cannot be a very high density of hydrogen in the stellar matter since at a great density the thermonuclear hydrogen reactions occur with great velocity and should release a great amount of energy. This is in contradiction to the weak luminosity of white dwarfs. Where do these stars obtain their energy? This is a singular question and we shall answer it later. At this point we will simply note that in the regions of stars where the stellar matter is composed of degenerate electron gas and the gas temperature of atomic residues is about 10^6–10^7 degrees there should be very little hydrogen, and in the calculation of the molecular weight it can be left out. On the other hand, in the calculation of the molecular weight of a degenerate electron gas, helium appears on equal terms with other heavy elements. In fact, the particular role of hydrogen and helium in calculations of the molecular weight was that in helium four units of atomic weight were distributed among three particles. In a degenerate gas the pressure depends only on the electrons. The role of helium nuclei under large pressures, as well as the role of other nuclei, is small. Consequently, four units of atomic weight of helium in a degenerate electron gas are distributed between two electrons formed during the fractioning of a helium atom. Thus, the atomic weight of the stellar matter composed of degenerate electron gas can always be considered equal to 2 (if there is no hydrogen).

Finally, why do we call degenerate electron gas a 'gas', notwithstanding the great density? The fact is that if the dimensions of the atoms are 10^{-8} cm then the 'dimensions' of electrons or bare nuclei are much smaller—about 10^{-12}–10^{-13} cm—and even at high densities in white dwarfs the distances between the particles in their stellar matter are much bigger than the dimensions of the particle.

We now introduce a simple formula describing the dependence of degenerate electron gas pressure on the density of the matter. Usually this formula is deduced using quantum mechanics. This was first realized by E. Fermi in the early twenties. We cannot give here the whole deduction, since we assume that the reader of this book does not know quantum mechanics. Nevertheless, it appears that this formula can be obtained very simply if we use the theory of dimensions. We shall describe this method in detail and to understand it we do not need to know quantum mechanics.

It is more convenient to find the dependence of the electron gas pressure p on the concentration of degenerate electrons n_e, i.e. on the number of these electrons in one unit of volume. The relation between the density of matter ρ and n_e is easily obtained, knowing that the molecular weight is, as already said, close to 2. We take $\rho = 2m_p n_e$, where m_p is the mass of a proton.

We now apply the method of dimension analysis. The pressure is measured in atmospheres or, in the CGS system, in units of dynes per cubic centimetres, i.e. in units of grams per centimetre per square second. The electron concentration is expressed by units per cubic centimetre. The pressure p depends not only on n_e but also on certain parameters defining the particularities of the matter. Since the degenerate electron gas is a phenomenon characteristic of quantum mechanics, its pressure should depend on the sole parameter which determines all quantum properties of the matter—the Planck constant $\hbar$. The numerical value and dimension of this constant in the CGS system is $\hbar = 10^{-27}$ ergs $= 10^{-27}$ g cm^{2}s^{-1}. We note that the gas pressure is an impulse transmitted from one gas particle to another and the surface which contains the gas. The impulse is determined by the mass of the particle, which in this case is the mass of an electron $m_e = 10^{-27}$ g. Electrons also have a charge e. However, the repulsion between the electrons is compensated for by their attraction towards the positively charged heavy ions situated among the electrons and therefore the electron charges have no influence on the electron gas pressure. In the conditions of a degenerate electron gas, the pressure does not depend on the temperature. As there are no other parameters we can consider that the pressure of a degenerate electron gas depends only on the values of n_e, $\hbar$, and m_e. The physical formulae describing this simple dependence always have a simple form—the product of certain degrees of all-important parameters. For this reason the dependence of pressure on the values of n_e, $\hbar$, and m_e will be

(22) $\qquad p = \Pi n_c^x \hbar^y m_e^z$

where Π is a certain dimensionless number of the order of unity, and x, y, z are degrees indicating how the pressure depends on the corresponding parameters. It is clear that the dimensions of the right- and left-hand sides of this equation should be the same. This requirement permits us to determine the values of the indices x, y, z, which is the main point of using the dimension method to find the physical formulae. We know the dimensions of all the parameters in formula (22). On the left-hand side we have gram (g) from p and on the right-hand side gram (g) from the product with a degree of $(y + z)$, since $\hbar$ as well as m_e have the dimension of gram (g). From this we obtain the equation $1 = y + z$. The dimension of units per centimetre (cm^{-1}) enters the left hand side while on the right-hand side this dimension appears twice: in $\hbar$ as square centimetres (cm^2) and in n_e as units per cubic centimetre (cm^{-3}). From the condition of equal dimensions we obtain a second equation: $-1 = 2y - 3x$. Finally, we have the dimension of units per square second (s^{-2}) from the left-hand side and on the right-hand side units per second (s^{-1}) appear only in the Planck constant. From this we obtain the equation $-2 = -y$ and immediately find $y = 2$. From the first equation we find $z = 1$, $y = -1$, and from the second equation $x = 1/3 \, (1 + 2y) = 5/3$. Thus, for the pressure of a degenerate electron gas we find the following formula:

$$(23) \qquad p = \Pi \, \frac{\hbar^2}{m_e} \, n_e^{5/3}.$$

The value of the dimensionless factor Π is not determined using the dimension method but this is not very important, since this value cannot be very different from unity. The most important fact here is that the dimension method permits the determination of the degrees for all important physical parameters in the formulae.

The numerical value of Π is obtained only from a complete calculation using quantum mechanics. It appears that $\Pi = 1/5(3\pi^2)^{2/3} = 1.9$. Introducing this value into (23) and replacing n_e by $\rho/2m_p$ we obtain the formula

$$(24) \qquad p = \frac{(3\pi^2)^{2/3}\hbar^2}{5m_e} \left(\frac{\rho}{2m_p} \right)^{5/3} = K\rho^{5/3}$$

where K is a constant, numerically equal to 3.1×10^9 atm cm^5/g$^{5/3}$. This formula also determines the basic characteristics of a degenerate electron gas. It is easy to understand that the pressure is proportional not to the first degree of the density but to a higher one. In fact, the pressure should increase as the density increases due to the fact that a great number of particles collide with the partition. In addition, as the density increases in a degenerate gas the number of electrons situated on the same trajectories, as well as their velocities, should also increase and consequently also the impulse transmitted by each electron. Thus the simultaneous increase of collisions with the partition and of electron velocity leads to a more rapid increase of pressure with the density.

This formula defines the degenerate gas pressure if the role of temperature is small. In particular, it is valid at the absolute zero of temperature. If the gas temperature (of electrons and atomic residues) is not zero, the possibility of applying it is determined by the relation between thermal velocities and the velocities which the electrons need to place themselves on a limited number of trajectories. If the thermal velocities are small—smaller than the velocities of degenerate electrons—the temperature can be neglected and formula (24) can be used, but if the thermal velocities are greater than those of electrons in a degenerate gas then, on the contrary, the characteristics of the gas are determined by its temperature and the degeneracy of electrons can be neglected. Then we must use formula (24). Instead of comparing the velocities we can compare the pressures: if the impulse transmitted to the partition by thermal motion is smaller than the impulse of electron velocities due to degeneracy (i.e. if the pressure according to (24) is greater than the pressure determined with Clapeyron's formula), then the gas will have the characteristics of a degenerate electron gas. In the opposite case the gas will not be degenerate. We write this condition in the form of an inequality:

$$K\rho^{5/3} > \frac{A}{2}\,\rho T.$$

A small transformation will give us the lower limit for the values of density at which the electron gas is degenerate:

$$(25) \qquad \rho > \left(\frac{AT}{2K}\right)^{3/2}.$$

We now introduce the numerical value: $\rho > (8.3 \times 10^7\,T/2 \times 3.1 \times 10^{12})^{3/2} = (T/75\,000)^{3/2}\,\text{g/cm}^3$. For example, at a room temperature of 300 degrees in the absolute scale, the electron gas becomes degenerate at a density of $2.5 \times 10^{-4}\,\text{g/cm}^3$ (!), i.e. the electron gas in our Earthly conditions is degenerate, e.g. in metals, where the density is of the order of $10\,\text{g/cm}^3$. Indeed, we must keep in mind that this formula is valid if $\mu = 2$. On Earth the molecular weight is much greater and the parameter K of formula (24) respectively smaller ($\mu^{5/3}$ times).

In stellar matter the temperature is, as we know, of the order of ten million degrees. One could imagine that the temperature in white dwarfs is of the same order. In fact these stars emit an energy, possibly due to thermonuclear reactions, for which a temperature of this magnitude is needed. We introduce into (25) a temperature $T = 14$ million degrees. We find that the stellar matter becomes degenerate if its density exceeds $1000\,\text{g/cm}^3$. The mean density of white dwarfs is tens and even hundreds of thousands of grams for $1\,\text{cm}^3$. Consequently white dwarfs should basically be composed of degenerate electron gas. Only the most exterior layers of the star where the density of the matter is less than $1000\,\text{g/cm}^3$ behave according to the ideal gas law. Here lies the solution of the paradoxal characteristics of the stellar matter of white dwarfs.

Thus at the high densities peculiar to white dwarfs, the pressure must be determined by formula (24) independently of the temperature. Therefore we cannot calculate the temperature in the centre of a white dwarf in the way it has been done in Chapter 2. Nevertheless, using this method we can calculate the density in the centre of a white dwarf, since the weight of the stellar matter column should be balanced at low temperatures necessary to counteract the great pressure like in ordinary stars, but a great density is also essential to counteract the great pressure in the degenerate gas of the white dwarf matter. Introducing into (3) the expressions for the pressure of the degenerate electron gas and ρ_c for the density in the centre of a white dwarf, we obtain

$$K\rho_c^{5/3} = 4f\,\frac{\bar{\rho}M}{R}$$

and the central density

$$(26) \quad \rho_c = \left(\frac{4f\bar{\rho}M}{KR}\right)^{2/5}.$$

Using this formula we calculate the density values in the centre of three white dwarfs given in Table 1. The results are shown in Table 4. The central densities of white dwarfs are four to ten times greater than their mean densities. Certainly at such densities a white dwarf is composed of degenerate gas not only at a temperature of several millions of degrees but also at temperatures of hundreds of millions of degrees. We underline again that the immense pressure in the centre of a white dwarf does not depend on temperature—a white dwarf can exist even at a temperature of absolute zero, when it will be not a white dwarf but a 'black dwarf', since at absolute zero temperature a star cannot radiate energy.

Table 4

Name of star	Central density, k/cm^3
40 Eridana B	400
Sirius B	250
Van Maanen 2	300

Thus the structure of a white dwarf does not depend on temperature and subsequently on luminosity. For this reason the mass–luminosity relation does not exist at all for white dwarfs. However, there is another relation which exists for white dwarfs; this is the mass–radius relation.

The necessity of this relation can easily be explained. In fact, we ask the following question: can white dwarfs of determined mass have different radii and thus different central densities? We admit this possibility. However, what will prevent them from passing from the state with a big radius to a state with a smaller radius? In an ordinary star this is not possible since in the 'attempt' of stellar contraction the temperature increases in the centre of the star and brings about a very strong increase of thermonuclear energy emission. On account of this the temperature and also the gas pressure increase even more, bringing the star back to its initial state.

Temperature is not essential for the equilibrium of a white dwarf. We can limit our consideration to the cool 'black dwarfs'. The stellar matter in all such dwarfs is the same (i.e. at the same zero temperature). We give a simple example: from pieces of the same metal with the same mass one can form compact spheres of only one radius. The same is true for 'black dwarfs'. From identical masses of the same stellar matter (degenerate electron gas at zero temperature) one can 'make' compact spheres of equal radii. If the temperature in the interiors of white dwarfs are different from zero, their radii (at identical mass) will also differ slightly but not by very much, since the temperature has no great influence on the pressure of a degenerate gas. We can compare this with the small growth of radius in the case of heated metal spheres.

Thus, the radius of a white dwarf is determined (in the first approximation) by its mass. Nevertheless, an important difference exists between white dwarfs and metal spheres (besides the disparity of masses). The radii of metal spheres made of one material are proportional to the cubic root of the mass, whereas the radii of white dwarfs decrease with the increase of mass. This phenomenon can be explained in the following way. In more massive white dwarfs the self gravitational force contracting the star is large. This leads to an increase in the central (and mean) stellar density and subsequently also to a decrease in the radius. The metal spheres do not fall apart as they are restrained by atomic forces which do not depend on the mass of the sphere; for this reason the density of the spheres does not depend on their mass.

The theoretical calculation of the mass–radius relation for white dwarfs is relatively complex but in the case of small densities this relation can be easily found.

Returning to formula (26), we can calculate that according to the data given in Table 4 the central densities in white dwarfs are four to ten times greater than their mean densities. A precise calculation shows that $p_c = 6\bar{p}$. A simple formula will give the relationship between the radius of the white dwarf and its mass:

$$(27) \quad R = \frac{3^{7/3}}{2^{5/3}\pi^{2/3}} \frac{K}{GM^{1/3}} \simeq \frac{2K}{GM^{1/3}} \simeq 8 \times 10^3 \left(\frac{M_\odot}{M}\right)^{1/3} \quad \text{km}$$

One can see that the radii of white dwarfs can be compared to the Earth's radius $(6.4 \times 10^3 \text{ km})$ and that they decrease as the mass increases.

Formula (27) is obtained in a very simple way and appears to be in good agreement with observational data if the mass of the white dwarf does not exceed half of the solar mass. For larger masses this dependence is more complex; therefore in Figure 8 we show the graph in a logarithmic scale of the

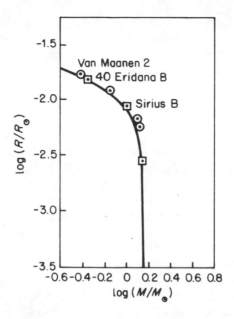

Figure 8 Mass–radius relation for white dwarfs

entire calculated dependence of a white dwarf's radius on its mass. The points indicate known white dwarfs (including those given in Tables 1 and 4). Observations confirm the theory quite well and prove that white dwarfs are effectively formed of a degenerate electron gas. Figure 8 also shows that at small masses the white dwarf's radius hardly varies. The left-hand part of the graph corresponds to formula (27) and is in good agreement with it.

This same figure shows that if the mass of the white dwarf approaches the limited $1.4M_\odot = 2.8 \times 10^{33}$ g its radius tends to zero, i.e. the self gravitational force of these stars is so big that the degenerate gas pressure is incapable of keeping the star in equilibrium. White dwarfs with a mass greater than $1.4M_\odot$ cannot exist in general. This is a very important theoretical result and we shall appreciate its meaning when studying stellar evolution.

Observations have shown that the mean of the white dwarf masses is 0.8–0.9 solar masses and their radii are on the average a hundred times smaller than the solar radius. Thus, 'ordinary', i.e. observed, white dwarfs are still far from the mass limit (often called the Chandrasekhar limit). What would happen if this limit is approached and what does the convergence of the radius to zero mean? Let us try to answer these questions.

We already know that an increase in the mass causes an increase in the central density. In its turn, the increase in density always leaves a smaller space for the elcctron trajectories. For this reason they must always occupy higher energetic states. In other words, the smaller the space for electrons the greater should be their energy. In the final account, at sufficiently high densities the electron velocities are close to the velocity of light. The motion of electrons will then obey the laws of the theory of relativity. Such a gas is called a degenerate relativistic electron gas. The equation of state is different, for it links the pressure and the density of the degenerate relativistic electron gas; it can also be deduced by the method of dimension analysis.

We must determine the relationship between the pressure p and the electron concentration n_e. As previously, this pressure is determined by the Pauli principle, i.e. it depends on the Planck constant $\hbar$. Contrary to the case of a non-relativistic gas, there is no dependence on the rest mass of the electron m_e. Since we now consider the case of relativistic electrons, their impulse is determined by the proximity of the velocity of the particles to the velocity of light c. Consequently, there should be a dependence of pressure on c. Instead of (22) we write

$$(28) \qquad p = \Pi n_e^x \hbar^y c^z$$

and with the same reasoning as for (24) we get $y = 1$, $z = 1$, and $x = 4/3$. Therefore the pressure of a degenerate relativistic gas is

$$(29) \qquad p = \Pi_1 \hbar c n_e^{4/3}.$$

The numerical value of a dimensionless constant is again determined only from the precise theory which yields $\Pi_1 = 1/4(3\pi^2)^{1/3} = 0.78$. Again replacing n_e by $\rho/2m_p$ we obtain

$$(30) \qquad p = \frac{(3\pi^2)^{1/3}}{4} \hbar c \left(\frac{\rho}{2m_p} \right)^{4/3} = K_1 \rho^{4/3}$$

where the constant K_1 equals 4×10^8 atm cm^4/g$^{4/3}$. With this equation we can also calculate the structure of a very dense white dwarf. This seems to be easy if we use the method that led us to formula (26), i.e. if we compare the gravitational pressure according to formula (3) with the pressure of the degenerate relativistic electron gas (30) calculated at the stellar centre:

$$(31) \qquad K_1 \rho_c^{4/3} = 4G \frac{\bar{\rho} M}{R}$$

However, later we meet a difficulty. In white dwarfs with an equation of state (24) the central density is six times bigger than the mean density. As soon as the electrons have velocities comparable to the velocity of light, the possibility of contraction of the degenerate gas increases and therefore in the central parts of white dwarfs with an equation of state (30) the density is much greater. In such stars the central density can be several tens of times greater than the mean density. Moreover, in this case the relation between the central and mean densities depends on the stellar mass: the greater the mass, the higher the electron velocities in the centre and the larger the central density. The precise theory shows that in the limiting case the central density is about fifty times larger than the mean density. In stars with such a large drop in density, formula (3) with factor 4 becomes inaccurate.

In order to obtain the formula we need we shall reason in the following way. The gravitational pressure (formula 3) is proportional to the product $G\bar{\rho}\, M/R \sim GM^2/R^4$. We do not know the proportionality coefficient and will therefore designate it by a, that is $p = aGM^2/R^4$. In very dense dwarfs the central density is proportional to the mean density $\rho_c \sim \bar{\rho} \sim M/R^3$. Here, too, we do not know the coefficient of proportionality and will designate it by the letter b. We assume that the values a and b are constant and independent of the mass and stellar radius. Then instead of (31) we have

$$K_1 \left(b\, \frac{M}{R^3} \right)^{4/3} = a\, \frac{GM^2}{R^4} \, .$$

We immediately see that the stellar radius falls out of this equation and its mass is clearly defined by

$$(32) \quad M = \frac{b^2}{a^{3/2}} \left(\frac{K_1}{G} \right)^{3/2} .$$

Thus, at first sight, we have obtained a strange result: in white dwarfs where the inner density is so high that the electrons move with a velocity approaching the velocity of light, the mass should be clearly defined and its radius in general does not depend on the mass. How can we understand this result? Let us go back to ordinary white dwarfs consisting of non-relativistic degenerate electron gas and do, at least mentally, the following experiment. We shall gradually increase the mass of the white dwarf by adding material from the exterior. This mental experiment does in fact happen in real conditions in cases where the white dwarf is part of a close binary system and the second star loses matter.

From Figure 8 and formula (27) we see that the mass of a white dwarf increases as its radius decreases and the density, especially in the centre, increases. In the central part the gas becomes relativistic, but according to formula (30) this brings about a certain slowing down of pressure growth as the density increases: if we had $p \sim \rho^{5/3}$ before, we now have $p \sim \rho^{4/3}$. As already noted, this means that the opposition of the gas to gravitational contraction will now be smaller and even a small stellar mass increase will bring about rapid contraction, especially in the central part. If we increase the mass a little more, a greater part of the stellar gas will become relativistic, the ability to resist contraction will decrease, and the stellar radius will rapidly decrease with the increase in mass. This corresponds to the sharp fall in the mass radius curve on the graph in Figure 8. Formula (32) also corresponds to the limiting mass of a white dwarf which can no longer resist the forces of gravitational contraction; for this reason the mass should not depend on the radius.

The exact theory gives the numerical values for the factors a and b. If, moreover, we keep the value for the molecular weight μ_e (without adjusting it to two as we have done above), then we obtain for the limiting mass of a white dwarf the following precise formula:

$$(33) \qquad M = 4.5 \left(\frac{K_1}{G} \right)^{3/2} \simeq \frac{3m_p}{\mu_e^2 (Gm_p^2/\hbar c)^{3/2}} \simeq \frac{5.75}{\mu_e^2} M_\odot.$$

The limiting mass of white dwarfs is called the Chandrasekhar limit. With $\mu_e = 2$ this limit equals $1.44 M_\odot$.

Formula (33) is interesting as it determines the characteristic mass of a star by universal constants. If the mass of a proton becomes greater, then the limiting mass of a white dwarf will decrease as m_p^{-2}. It has been assumed that the gravitation constant G varies with time; then the limiting mass of white dwarfs should also vary as $G^{-3/2}$. Indeed, the gravitation constant does not change and the magnitude of the mass of a proton is fixed by the value $m_p = 1.64 \times 10^{-24}$ g, although we do not know why this is really so. The relation of astronomical parameters and microscopic parameters described by formula (33) is always of continuous interest.

Nevertheless, at very large densities we must consider not only the special theory of relativity but also the general relativity theory, according to which in the proximity of massive bodies the characteristics of space and time change. It is beyond the scope of this book to explain the general theory of relativity (the reader will find many books on this subject) but we shall explain a few consequences. Close to the surface of a body with a large gravitational force, space is distorted and time changes. This change of the curvature of space can be observed by the deviation of the light beam at an angle equal to about $4fM/c^2R$ radians and after a decrease in the velocity of light to fM/cR centimetres per second. Here M is the mass of the body, R is its radius, and c the velocity of light in the vacuum. The larger the mass and the smaller the radius, the greater are the effects of the general theory of relativity.

We shall now study the contraction of a white dwarf taking into account the general theory of relativity. The decrease in the velocity of light and the increase in the curvature of space prevents the electrons from occupying higher energy levels and occupying a smaller volume—the curved space has, at a given dimension, a smaller volume than the 'plane' space. As a result the counterpressure against contraction decreases and at a density higher than a certain critical value the white dwarf ceases to be stable, i.e. can no longer exist. The magnitude of the density limit of the most massive white dwarfs is variable, depending on the hypothesis of the chemical composition of the matter of white dwarfs. If a white dwarf is composed of helium, the limiting density will be about 10^9 g/cm^3. However, if the majority of nuclei of the white dwarf's matter is composed of iron (this is the upper limit of formation of chemical elements in stars) then the limiting density increases by up to 2×10^{10} g/m,^{3}. The radius of such a star will be about 1000 km—six times smaller than that of our Earth.

We have still not arrived at the limit of the possible. At very large densities the process of the so-called neutronization of matter starts. Its nature can be presented as follows. The atomic nucleus can absorb an electron which will transform one of the protons of the nucleus into a neutron. In this process one neutrino will be released, which will freely leave the star and carry away part of the energy that belonged to the absorbed electron. In general conditions the nucleus formed with an excess of neutrons is unstable and in the process of radioactive decay it should again return to the initial nucleus, releasing a new electron. In a white dwarf this is not possible—a re-expelled electron has less energy than the initial one and there is no place for it in the degenerate gas. The nucleus thus remains with an excess of neutrons. This process continues until the nuclei absorb the greatest part of the electrons and are converted into 'neutron gas'.

Such neutronization starts at densities smaller than the limiting value defined by the general theory of relativity (i.e. at $p_c \sim 10^9$–10^{10} g/cm^3), and at nuclear densities of about 10^{14}–10^{15} g/cm^3 practically the entire gas becomes neutron gas. This leads to the formation of neutron stars, which we shall study in the next chapter. Here we only note that on account of neutronization of matter the mass limit of white dwarfs decreases slightly—instead of 1.4 of the solar mass this limit will be about 1.2 of the solar mass.

Until now we have studied the structure of a white dwarf, only considering 'black dwarfs' with reference to their temperature. We did not pay attention to the fact that they radiate energy. For this reason we shall now study energy sources and the energy transfer in the interior of these stars.

As we know, the luminosity of an ordinary star is determined by the transparency of the stellar matter. How transparent is the matter of a white dwarf? As strange as it would appear at first sight, the matter of a white dwarf notwithstanding the great density, is quite transparent. This phenomenon can be easily explained. The reason for the opacity of ordinary stellar matter is the absorption of light by electron transitions from a close orbit to a remote one. The remote orbit should therefore be free of electrons or have only one electron. In fact we know that according to Pauli's principle one orbit should contain no more than two electrons. This is also true for free electrons in the stellar matter: in order to absorb luminous energy they must pass from one trajectory to another. However, in the degenerate matter of a white dwarf all trajectories are occupied, the electrons can go nowhere, and therefore they cannot absorb energy. It is true that some electrons with high energies can nevertheless pass to another trajectory if they can increase their velocity to create a difference from the velocity of the electrons which already occupy that trajectory. Thus, light can be absorbed in a degenerate gas only by a few rapid electrons and consequently the transparency of a degenerate electron gas is comparatively high. The stellar matter of white dwarfs has a good thermal conductivity. This is testified by the well-known fact that metals, which are also composed of degenerate electron gas, are good conductors of heat and electricity.

Thanks to the high transparency and thermal conductivity of the degenerate electron gas, the temperature in the interior of white dwarfs is nearly constant. In fact, good thermal conductivity rapidly smooths out the temperature. Of course we must take into account the fact that a white dwarf is not entirely composed of degenerate gas, since the degeneration takes place only at densities exceeding $1000 \, g/cm^3$ (at a temperature of 14 million degrees). For this reason the external layers of a white dwarf are composed of ordinary, not degenerate, ideal gas with poor transparency, as in any 'ordinary' stellar matter. This layer of 'non-degenerate' stellar matter detains the thermal flux coming from the interior of the white dwarf. In an ordinary star the temperature increases from thousands to ten millions of degrees over the length of its radius. In a white dwarf the temperature increases from thousands to ten millions of degrees in a narrow surface layer with a thickness ten times smaller than its already very small radius, and then for the rest of the distance to the centre it remains nearly constant.

We can describe the structure of a white dwarf as an immense sphere composed of degenerate electron gas with a density in the centre of hundreds and thousands of kilograms for one cubic centimetre. The temperature in the interior of the sphere is almost constant. On the outside it is surrounded by a comparatively thin layer (a few per cent. of the radius) of ordinary stellar matter, where the density falls from thousands of grams per cubic centimetre to zero and the temperature decreases from millions to one thousand degrees.

In the outer gas envelope there should exist an energy flux towards the outside. Consequently, either in the degenerate gas or on its limit there should be sources of energy. We already know that there is no hydrogen in a degenerate gas: at a high density it would enter into a reaction releasing a great amount of energy and this is not observed. Calculations show that in the degenerate part of a white dwarf the hydrogen abundance cannot exceed 0.05 per cent. On the other hand, observations show that hydrogen exists in a gas envelope, at least in its surface layers (this was discovered through spectral analysis). Therefore we can make an essential assumption: thermonuclear reactions in a white dwarf can occur in a gas envelope in a thin layer immediately adjacent to the degenerate stellar matter. The energy releasing layer must be effectively thin because of the strong dependency of thermonuclear energy release on temperature. The fact that this layer is thin explains why so little energy is released in white dwarfs although the temperature is the same as in ordinary stars. Another question arises: why is there no hydrogen in a degenerate gas when it has been conserved in the gas envelope? This will be answered in our study of stellar evolution. Indeed, theory shows that such an energy release is unstable in the close surface layers. The white dwarf should in this case pulsate and this has not been confirmed by observation.

Even if there is no hydrogen in the gas envelope of a white dwarf, it can nevertheless radiate as a result of cooling. In fact in the interior of the white dwarf there is a supply of thermal energy which slowly filters through the opaque, isolating gas envelope. Since this filtering process is slow the cooling of the white dwarf takes a long time. It is easy to calculate the duration of cooling of a white dwarf.

However, we must first determine the temperature in the degenerate part of a white dwarf. It is evident that the temperature of a white dwarf depends only on its luminosity—as a matter of fact, we already know that the pressure in the interior of a white dwarf does not depend on the temperature and consequently its radius and mass are not related to temperature either. It is quite simple to deduce the formula for determining the interior temperature of a white dwarf from its luminosity, but we shall give only the final result:

$$(34) \qquad T_6 = 50 \left(\frac{L}{L_\odot} \frac{M_\odot}{M} \right)^{2/7} \qquad \text{million degrees}$$

If we know the luminosity and mass values of the white dwarf we can easily find its interior temperature with the help of this simple formula.

Now let us see how to calculate the cooling of the white dwarf. It is known that the amount of heat contained in 1 g of a single atom gas equals $3/2 \, AT/\mu$ (A is the gas constant and μ the molecular weight of the gas composed at atomic residues; μ should not be confused with the molecular weight of one electron μ_e which is always close to 2 as the value of μ can be much greater). The same amount of thermal energy is also in 1 g of degenerate electron gas where it is distributed over the gas of 'atomic residues'. This is the so-called 'storehouse' of thermal energy in a white dwarf. Now we can express the total supply of thermal energy in a white dwarf with the formula

$$(35) \qquad W = \frac{3}{2} \frac{AT}{\mu} M$$

where T is the temperature in the degenerate part of the white dwarf and M is its mass. Since the gas layer represents a few per cent. of the stellar mass, M can be considered as the mass of the whole white dwarf. It is evident that if we divide the energy supply of the white dwarf by its luminosity, i.e. by the amount of energy radiated in 1 s, we obtain the cooling time of the white dwarf:

$$(36) \qquad t = \frac{100}{\mu} \left(\frac{L_\odot}{L} \frac{M}{M_\odot} \right)^{3/7} \qquad \text{million years}$$

This formula gives a lower limit for the cooling time since as time passes the luminosity of the white dwarf decreases.

The cooling time also depends on the chemical composition of the white dwarf. We already know that there is no hydrogen but it is difficult to determine the amount of helium and other elements. If in the evolution process of a star, which in the end turns into a white dwarf, the temperature has not reached a hundred million degrees then the entire white dwarf will probably be

Table 5

Name of star	Temperature in millions of degrees	Cooling time of years
40 Eridana B	13	400
Sirius B	9	190
Van Maanen 2	700	1200

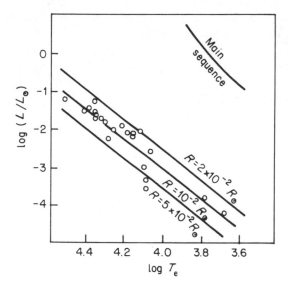

Figure 9 The temperature–luminosity diagram for white dwarfs

composed of helium ($\mu = 4$). This is probably the case for stars with a small mass. In bigger stars the temperature could be higher during this evolution process and therefore white dwarfs can exist which are composed of magnesium and even iron ($\mu = 24$ and $\mu = 56$ respectively). In the latter case the white dwarfs cool more quickly. Considering that the known white dwarfs are of small mass we shall limit ourselves to those composed of pure helium. Using the formula given above we obtain their temperatures and characteristic cooling times which are given in Tables 5. This shows us that, first, the temperatures of white dwarfs are relatively small and correspond entirely to ordinary temperatures in which thermonuclear reactions take place and, second, if thermonuclear reactions are not sufficient (as, for example, in the case of Van Maanen's star) the cooling of the white dwarf can assure its luminosity during hundreds of millions of years. This concerns the problem of stellar evolution which will be considered in Chapter 10.

To conclude this chapter we can now show on the lower left-hand part of the Hertzsprung–Russel diagram the positions of white dwarfs (Figure 9). On the abscissa, in place of the spectral type we have the corresponding surface temperature. Almost all white dwarfs have radii within the range of $0.02R_\odot = 1.4 \times 10^9$ cm to $0.005\,R_\odot = 3.5 \times 10^8$ cm. The cooling of white dwarfs is indicated on the graph by points parallel to the lines of equal radius down to the right-hand side.

Thus all particularities of white dwarfs (high densities, small radii, low luminosity) have been entirely explained by modern physics. It is difficult at first sight to expect an analogy between the behaviour of free electrons in a metal well known for electron conductivity and the unusually dense matter of white dwarfs. The physical nature of one and the phenomenon of the other is the same. The difference lies only in the scales.

6

Superdense stars—pulsars and 'black holes'

In the preceding chapter we saw that if the dense matter of a white dwarf is continuously contracted this matter will be neutronized—the electrons are 'crushed' into the atomic nuclei, react with protons, and turn into neutrons. Such nuclei will then disintegrate and neutron gas appears. One could imagine, at least in principle, that the star is contracted to a state where all the matter is converted into neutron gas. We shall call such a star a neutron star. However, it must be noted that real neutron stars which are observed, such as pulsars, have a more complex structure. There is neutron gas as well as hyperon gas, ordinary plasma, and hard crust. In this new astrophysical problem we shall proceed in the following way. First we shall study a simple case of a star entirely composed of neutron gas. Then we shall describe a real model of a pulsar. After this, we shall see how they are formed. Finally we shall consider denser objects—'black holes'.

Let us imagine a sphere composed of pure neutron gas held together by the self gravitational force. How can we determine the parameters of such a sphere? How, for example, can we find its radius if the mass of the sphere is given?

At low temperatures neutron gas has the same behaviour as an electron gas—in this case quantum degeneration also appears. The greater the amount of neutrons in a limited volume, the greater their energy and even greater the pressure of the neutron gas. For this reason the calculation of the neutron gas pressure can be done in the same way as for the degenerate electron gas. Here the pressure p depends only on the concentration of neutrons n_n, the Planck constant $\hbar$, and in this case on the mass of the neutron m_n. Using the method of dimension analysis we obtain the same formula as (22) with m_e replaced by m_n. The neutron gas pressure is

$$(37) \qquad p = \Pi \, \frac{\hbar^2}{m_n} \, n_n^{5/3}.$$

Here also the value of Π is not determined by considering the dimensions. We shall take the same value as that used for an electron gas, i.e. we assume $\Pi = 1.9 \approx 2$. The molecular weight of the neutron gas is very close to unity since the masses of a proton and a neutron are almost equal. Therefore we obtain for the neutron gas pressure

$$p \approx 2 \frac{\hbar^2}{m_n} \left(\frac{\rho}{m_n} \right)^{5/3} = \frac{2\hbar^2}{m_n^{8/3}} \rho^{5/3} = K_n \rho^{5/3}$$

Where the constant $K = 5.3 \times 10^3$ atm cm^5/g$^{5/3}$. If we ignore the difference between molecular weights, the difference between the degenerate electron and degenerate neutron gas at equal densities will be of a factor equal to the mass ratio of a neutron and an electron.

Unfortunately formulae (37) and (38) are not very precise. This is due to the fact that in a neutron gas the density is quite high, which means that the neutrons are close to one another. In this case the pressure of the neutron gas is affected by the strong mutual attraction of the neutrons at small distances, i.e. the action of nuclear forces appears, which in general retains the neutrons in the nuclei. However, at extremely short distances the neutrons are repelled. In order to calculate the nuclear attraction and the repulsion of neutrons Cameron assumed the following formula for the pressure of a neutron gas:

$$p = 5.3 \times 10^3 \rho^{5/3} + 1.6 \times 10^{-11} \rho^{5/3} - 0.14 \rho^2 \qquad \text{atm.}$$

Here the second term takes into account the repulsion at small distances—it becomes larger at very high densities—and the third term takes into account the mutual attraction of neutrons which evidently decreases the pressure of neutron gas proportionally to the square of the number of particles. But in the first approximation we can still use the simple formula (38).

Knowing the equation of state of a neutron gas we can now consider neutron stars. Indeed, we can employ methods entirely analogous to the theory of the structure of white dwarfs. Here formula (27) is valid, only now in place of the 'electron' value of the constant K we must write the 'neutron' value for this parameter. We then obtain

$$(39) \qquad R \approx \frac{2K_n}{GM^{1/3}} \approx 12 \left(\frac{M_\odot}{M} \right)^{1/3} \qquad \text{km.}$$

The radius of neutron stars is only a few kilometers! This can be seen from even simpler estimations. The gas will be almost entirely composed of neutrons only if its density is of the order of 10^{14}–10^{15} g/cm^3. The radius of a homogeneous sphere with a mass equal to the mass of the Sun and a density of 3×10^{14} g/cm^3 is $R = (3M_\odot/(4\pi\rho))^{1/3} = 12$ km, which is a value close to (39). Certainly the density of a neutron star is not homogeneous, but it is evident that neutron stars should always have a small radius.

Thus a simple model of a neutron star composed of pure neutron gas describes a configuration with a mass approximately equal to that of the Sun and with a radius of 10–12 km. As in white dwarfs, the central density is six times greater than the mean density. In reality, however, the model of a neutron star is more complex. The matter cannot remain a neutron gas up to its surface. Therefore we must consider the structure of a neutron star taking into account all possible changes in the state of the matter. After the discovery of pulsars, astronomers and physicists paid particular attention to neutron stars and many papers were devoted to a theoretical analysis of neutron star structure. We shall now briefly expose the results of these studies.

First of all, let us note that a neutron star is formed from a hot star after a strong contraction. It is therefore evident that the temperature in the interior of a neutron star will remain high. There are no precise data, but probably the surface temperature of neutron stars is about hundreds of thousands of degrees and in their innermost the temperature reaches hundreds of millions of degrees. Therefore the inner temperature close to the surface of neutron stars should be relatively small compared with those in the deeper layers which increase to very high values.

At the very surface of a neutron star the density is about 10^4 g/cm^3. This is the density of the matter of white dwarfs composed of atomic nuclei and degenerate electrons. It is true that here the matter is in a very strong gravitation field of the neutron star and moreover its temperature is very much smaller than in the interior of white dwarfs. Therefore this matter has the characteristics not of a gas but of a hard body.

The chemical composition of the outer layers of neutron stars is apparently such that the most common element is iron, although there can also be atomic nuclei of other elements. In the atomic nuclei of iron, protons and neutrons are densely packed. These atomic nuclei are the final product of all nuclear reactions with energy release. Thus, the exterior part of neutron stars represents a hard, primarily iron, crust where the density increases from the surface towards the interior. It is thought that 'mountains' exist on the very surface of the crust of neutron stars, as on the Earth's surface. However, the height of these 'mountains' does not exceed several tens of centimetres as higher 'mountains' are not possible due to the great force of gravity.

The hard crust of neutron stars composed of ordinary atomic nuclei and degenerate electron gas has a thickness of about 100 m, reaching a depth where the density of matter is 4.3×10^{11} g/cm^3. In neutron stars with a small mass $(0.13 M_\odot.)$ the hard crust can occupy half of the radius of the star, but in the 'heaviest' neutron stars its thickness is not more than 100 m. If we go deeper into this crust we note that there is a slight change of the chemical composition. Close to the surface iron prevails with an atomic weight of about 56, but close to the inner limit of the hard crust there exist more atomic nuclei of the type of zirconium (with the atomic number 40) and of nuclei with very large atomic weights reaching 127. This is due to the fact that the electrons 'crushed' into atomic nuclei decrease their atomic number but increase the number of neutrons in these nuclei. The decrease of the Coulomb repulsion force favours the fusion of nuclei and also a total conversion into neutron gas.

The most interesting layer is the outermost one, several meters thick, where the density of matter does not exceed 10^5 g/cm^3. This is also a hard layer with the particularity that the material of the neutron star's crust is strongly magnetized. The magnitude of the magnetic field reaches 10^{10}–10^{12} G. We shall study these magnetic fields later.

The intense transformation of matter composed of atomic nuclei and degenerate electrons into a neutron gas starts at depths where the density exceeds 4.3×10^{11} g/cm^3 and even at depths where the density reaches 10^{12} g/cm^3 there is a greater amount of matter with free neutrons than with atomic nuclei. At depths where the density reaches a magnitude of 3×10^{14} g/cm^3 almost all of the matter has become neutron gas.

The thickness of this neutronization layer is about 100 m in the heaviest neutron stars. Thus, the entire thickness of the crust of ordinary neutron matter of a neutron star, having about the same mass as the Sun, represents 200 m—and this for a total stellar radius of about 10 km! Here the largest part of the neutron star matter is in the form of neutron gas and therefore we can apply exactly the same approximations as used previously. In not very massive neutron stars (e.g. at $M = 0.1M_\odot$) a density of 3×10^{14} g/cm^3 exists only close to the centre, since the mass is too small to let the star be contracted to greater densities. Here the stellar matter represents a mixture of neutron gas, atomic nuclei, and electrons. The study of the structure of such stars is complex.

From this we can draw an important conclusion. Neutron stars cannot exist if their mass is small: in such stars the density is small and the neutrons disintegrate into protons and electrons. Until now it has not been possible to calculate precisely the limiting mass of neutron stars, but it should not be less that $0.1M_\odot$.

At a density of $\rho \geq 3 \times 10^{14}$ g/cm^3 the neutron star matter is composed of neutrons with a small addition of protons and electrons (4 per cent. of each charged component). The pressure is almost entirely determined by the density of neutrons. At even greater densities ($\rho \geq 10^{15}$ g/cm^3) in the central parts of most massive neutron stars there are also hyperons, elementary, charged, and neutral particles with masses greater than the mass of a proton and a neutron. It is probable that the neutron gas behaves like a superfluid and the pressure of the charged particles adds the characteristics of a superconductor to this matter.

The study of neutron stars takes into account all particularities of the behaviour of the matter at great densities. However, the general dependence of the radius of a neutron star on its mass, described by formula (39), will still be valid if the mass of the neutron star is not too small.

We already have mentioned that neutron stars are pulsars. The fundamental characteristic of a pulsar is that distinct impulses of radiation are observed with a well-defined periodicity. How can this periodicity be explained? It is evident that this can be done through either one of two phenomena: pulsations, i.e. periodic contraction and expansion, or rotation.

Pulsating variable stars composed of ordinary gas will be considered in Chapter 8. There we shall also study the causes of periodic expansion and contraction of stars. It is true that these causes cannot be applied to neutron stars but we shall assume that they can also pulsate. In Chapter 8 we obtain relation (46) which determines the pulsation period for a given mean density. This relation is universal and can be applied to neutron stars as well as to ordinary gas stars. We introduce into formula (46) the value for the mean density $\bar{\rho}$ equal to 3×10^{14} g/cm^3 and find that if a neutron star starts pulsating the pulsation period should be about 7×10^{-4} s. Even the shortest periodic pulsar has a period of 3×10^{-2} s. Thus the pulsations cannot explain the phenomenon of the periodicity in the radiation of pulsars and moreover one can consider that neutron stars in general do not pulsate.

It remains to assume that they rotate and that the period of impulse repetition is the period of rotation. A pulsar with a period of 3.3×10^{-2} s makes 33 r/s and even the slowest rotating pulsar, with a period of 4.8 s, makes 0.21 r/s.

Such rapid rotations can be due to the origin of pulsars. We assume that initially a pulsar was a big gas sphere with radius R_1, rotating with a certain period P_1. In fact, rotating stars are observed and the rotation periods of certain stars can even be about 30 hours, although the rotation period of the Sun is much greater—almost 30 days. We assume that in its evolutionary course this star was contracted to the state of a neutron star. Everybody knows that in this case it should rapidly untwist. The variation in the rotation state is determined by the conservation law of the rotational momentum. As we know, the rotational momentum is the product of the moment of inertia I and the angular velocity of rotation, equal to $2\pi/P$. In order to calculate the moment of inertia we must know the distribution inside the star, but I will aproximately be $0.1MR^2$, where M is the stellar mass and R the radius. Consequently, at the contraction of the star the product is $I \times 2\pi/P = 0.6MR^2/P$. If the mass does not change the value R^2/P stays the same. A pulsar with radius R_2 generated from a star with radius R_1 will rotate with a period P_2, expressed by the formula

$$P_2 = \left(\frac{R_2}{R_1}\right)^2 P_1.$$

Let us now consider some numerical values. Let the stellar radius be equal to 7×10^5 km, up to the contraction, that is the same as the radius of the Sun. If we suppose that the radius of a neutron star is 14 km, we find that after the contraction the rotation period should decrease by about 2.5×10^9 times. A neutron star originating from the Sun will make one turn in 10^{-3} s. It is possible that the initial rotation periods of neutron stars were really as short as this and that the presently observed pulsars have slowed down their rotations.

Deceleration of rotation exists in pulsars. In many of them, particularly in those with small periods, there is in fact a gradual but very regular increase of the period. As a rule, the smaller the period, the more rapid its increase. In the most rapidly rotating pulsar with a period of 3.3×10^{-2} s, this value will increase by 1.3×10^{-5} s in a year. In another rapidly rotating pulsar with a period of 8.9×10^{-2} s, the increase of the period will be 3.5 times slower, and for other pulsars it is even more.

In two of the most rapidly rotating pulsars clitches were observed in the rotation when the period suddenly decreased with a jump and then continued to grow. The clitches in the period were small and can be explained by the small variations in the radius of the neutron star or by the redistribution of matter in its surface layers. There is a theory that flows and 'starquakes' can occur in the crust of a neutron star which give rise to clitches. However, other theories also exist and the phenomenon of these clitches is still not very clear. Moreover these events are quite rare.

Observations show that neutron stars decelerate their rotation, i.e. they lose rotational energy. The value of the rotational energy of a neutron star is $\frac{1}{2} I(2 \pi/1P)^2$ and knowing the rotational deceleration we can easily determine how much energy a pulsar should lose in a unit of time. For example, in the most rapidly rotating pulsar the rotation energy equals 5×10^{48} erg. The change of period by 1.3×10^{-5} s in a year corresponds to an energy loss of about 10^{38} erg/s. Of course, in other more slowly rotating pulsars, the rotation energy losses are much smaller but the energy radiated by pulsars in one second can be much greater than the luminosity of the Sun due to the deceleration of rotation.

Let us now consider the causes of rotation deceleration in pulsars. The most effective cause appears to be the magnetic field, although maybe there are others. We shall now study in detail the magnetic field of a pulsar.

First of all, how can such a magnetic field appear? It is known that at least some stars have magnetic fields. These fields can be weak, as in the Sun, or much stronger (reaching 35 thousand gauss). Suppose that a star, together with its magnetic field, contracts. In this case the magnetic lines of force are also contracted, meaning that the magnetic field grows. One can show that the strength of the magnetic field is greatly increased as the rotation period decreases. Therefore, if the Sun were to contract to the dimension of a pulsar its magnetic field would have a strength of 2.5×10^9 G. A star with an initial magnetic field of 10^3 G would generate a pulsar with a magnetic field tension of 10^{12} G.

Therefore, a pulsar formed after a stellar contraction rotates rapidly and has a strong magnetic field. It should lose energy by radiation in the same way as electric charges lose energy by rotation on a certain orbit with a high velocity. The magnitude of energy loss is determined by the following formula:

$$(40) \qquad L \approx \frac{H^2 R^6}{c^3} \left(\frac{2\pi}{P} \right)^4 .$$

Here H is the magnetic field strength at the surface of the pulsar, R is the radius, c the velocity of light, and P the rotation period. In formula (40) the value L is known from observations of the variation of the period ($L \leq 10^{38}$ erg/s), the radius of the star being taken as equal to 12 km. Introducing these values into formula (40) and taking the rotation period to be 3.3×10^{-2} s, we find that the magnetic field strength for this star is equal to 12^{12} G. A pulsar with a magnetic field, in fact, slows down its rotation by radiation of electromagnetic energy, which is indeed observed. It is easy to explain the origin of the strong magnetic field and the rapid rotation of the pulsar.

On the other hand, the same magnetic field and rotation explain the particular character of their radiation which we observe as separate impulses. As this book treats the inner stellar structure of a pulser we shall not examine the theory of radio emission which is due to the hard crust of the pulsar. We shall only make a few remarks here.

The magnetic axis of a pulsar does not coincide with its axis of rotation. Therefore as the pulsar rotates, its magnetic axis draws a cone in space. The magnetic field of the pulsar is strongest in the vicinity of the magnetic poles, where this axis leaves the surface of the pulsar. Here the magnetic field is almost perpendicular to the surface. We can assume that electromagnetic radiation is emitted by regions with the strongest magnetic field and that this radiation comes out of these regions principally along the magnetic lines of force. In other words, it is quite probable that the pulsar generates radio emission only along its magnetic axis. Therefore we observe the radiation of a pulsar only when we are in the prolongation of its magnetic axis. This explains the pulsating character of the radiation of a pulsar—it is simply the displacement of the ray of the pulsar over the sky. The observation of the radiation of a pulsar resembles the observation of light from a lighthouse.

Let us come back to the structure of neutron stars. We have already seen that their mass cannot be smaller than 0.1 of the solar mass. Is there an upper limit for the mass of neutron stars? It appears that the masses of neutron stars are also limited from above. Again we turn to the analogy between degenerate electron stars (white dwarfs) and degenerate neutron stars (pulsars). Here the bigger the mass of the star, the smaller its radius, the denser the matter, and the greater the velocities of neutrons. Therefore, we can expect that as the mass of a neutron star increases the neutron gas becomes relativistic, and instead of the equations of state, (37) and (38), we can use the equation of state (30) which does not depend on the mass of the particles. There will also be a mass limit, as was the case with the Chandrasekhar mass limit in white dwarfs. For the neutrons to become relativistic the neutron gas density should greatly exceed the nuclear density of the matter. This is therefore probably not the effect that determines the mass limit of neutron stars.

We now shall imagine what would happen if we increased the mass of neutron stars. The greater the mass, the smaller the radius and the greater the velocity necessary for the particle to leave the stellar surface. We will use the well-known formula which permits us to find the velocity that a body needs to leave the surface of the Earth:

$$(41) \qquad \frac{1}{2} v_k^2 = \frac{GM}{R}$$

We obtain this formula by comparing the kinetic energy of a body $\frac{1}{2} mv_k^2$ to its potential energy GMm/R. Formula (41) can also be applied to a pulsar. On Earth the velocity v_k is not very high ($v_k \approx 11.2$ km/s) but on a pulsar v_k almost reaches the velocity of light.

If we continue to increase the mass and decrease the radius of a neutron star we can obtain from (41) a velocity v_k equal to that of light. From this moment the star 'retires within itself'—i.e. even light can no longer leave its surface. For this reason physicists and astronomers say that the star recedes to its 'black hole'. This name expresses very well the following phenomenon: light or particles can be attracted by this body but nothing leaves it for outer space!

The radius at which the velocity of departure becomes equal to the velocity of light is called the gravitational radius:

$$(42) \qquad R_g = \frac{2GM}{c^2} = 3 \frac{M}{M_\odot} \qquad \text{km}$$

where R_g is determined by the mass of the body. The star can be in a stationary state and can only be seen if the real radius is larger than the gravitational radius. What would happen if the radius is smaller than its gravitational radius is hard to tell, but this is not so important. However, no radiation leaves the star and so we shall never know anything about them. In the general theory of relativity the existence of 'black holes' is related to the curvature of space–time. If the star has a radius smaller than its gravitational radius the space will close upon itself. Of course the star does not disappear into its 'black hole' without trace: the gravitational field of the star in its 'black hole' remains and it continues to attract the surrounding matter and to pull it into the 'black hole', gradually increasing its mass. Collisions can bring about a fusion of 'black holes' into one big 'black hole'.

The radii of neutron stars, as follows from formula (39), are dangerously close to their gravitational radii and this effect determines the mass limit of stationary neutron stars. If we compare the radius of a neutron star from (39) with its gravitational radius (42) we obtain the mass limit:

$$(43) \qquad M \leqslant \left(\frac{K_n c^2}{G^2} \right)^{3/4} \approx 2.8 M_\odot.$$

Although this calculation is approximate, an even more precise theory leads to almost the same value.

The masses of neutron stars are limited from above by a comparatively small value—not more than about three solar masses. This is an important conclusion. If a star with a mass exceeding this limit starts to contract this contraction does not stop with either the degenerate electron gas or the degenerate neutron gas; the contraction does not come to a stop at all and the star recedes into its 'black hole'. This deduction is very important for the theory of stellar evolution and we shall study it in more detail in Chapter 10.

Theoretically the existence of 'black holes' was predicted by Oppenheimer and Snayder in 1939 but the real proof of their existence came in 1972–1973 with the demonstration that the X–ray source Cyg X–1, a close binary system with a rotation period of 56 days, has a mass of about 10–12 solar masses. Thus we have here a 'black hole' which attracts the matter. When falling into the 'black hole' the matter is heated and emits X-rays. Obviously 'black holes' can in general be 'observed' only if they are part of a close binary system.

7

Calculation of stellar models

In physics and astrophysics, as in many other fields of science, one often uses so-called models of diverse phenomena. By the term 'model' we do not mean a mock-up made of wire, paper, or some other material; 'model' here means a calculation, a numerical description of the corresponding physical object or phenomenon. If, on the basis of a calculation, we compose a table or a graph which gives the values of density, temperature, and chemical composition at different distances from the centre and which shows how the properties of stellar matter change at different depths, we have calculated a model of the star. Why do we say that we calculate a stellar model and not describe a star? In fact, in constructing a model we do not take into account all physical factors but only the more important ones. The model is not a precise copy; it describes only the fundamental particularities of the object. Models of stars do not correspond to their real structure in all details— we do not know them and probably never will. We do not need all the details; our task is to show the basic rules which define the laws of stellar structure and its evolution. This can be fully realized if we build a more or less satisfactory stellar model describing only its basic properties.

Stellar modelling using calculations is justified as we cannot observe the conditions in stellar interiors. The calculation of a model based on physical laws allows us to understand the physics of stars. As a matter of fact, the whole theory of inner stellar structure amounts to the calculation or, as is often said, the construction of different stellar models, their comparison, and an analysis of differences between one model and another. To build models one starts from the physical laws described in the preceding chapters and from optical data of stellar masses, luminosities, and radii.

We already know the physical laws used by astronomers in the theory of stellar model calculations in the form of a system of comparatively simple differential equations which we shall not give here. These laws are the following:

1. The equilibrium between the gravitational force of the star and the gas pressure of the stellar matter (Chapter 2)

2. The transfer of energy from the interior to the exterior through the opaque stellar matter (Chapter 3)
3. The law of energy emission in thermonuclear reactions (Chapter 4)
4. The equation of state of an ordinary gas (Clapeyron's equation in Chapter 2) or the equation of degenerate electron gas in white dwarfs (in Chapter 5)
5. The determination of molecular weight and chemical composition of the stellar matter (Chapter 2)
6. The relationship between the stellar matter opacity and the temperature and density (Chapter 3)

The basic parameters of stars (their mass, luminosity, and radius) are also known (Chapter 1). The author wishes to emphasize that the calculations of stellar models are based on the physical laws studied in this book. The reader knows many of them from the school physics course. The differential and integral calculations used by astronomers are a matter of technique and methods of computation and not of the physics of theoretical principles.

Stellar models have been calculated since the twenties. In the beginning they were calculated by hand with slide rules. The calculation of one model needed months of hard work. Now the models are calculated on powerful computers and one model is obtained within a few minutes or even seconds of computer time. Presently thousands of models are being calculated from which a large number of important deductions on the structure and evolution of stars can be made.

In the first editions of this book we gave in detail the methods used to calculate stellar models 'by hand'. Now this is no longer of use. Although the principles of computer calculations are analogous to those of hand calculations it is more convenient to use here an iterative method which is simpler to explain.

Let us assume that for a given stellar model we know how the density and gas temperature depend on the distance from the stellar centre r, in other words, we consider that the functions $\rho\,(r)$ and $T(r)$ are known even if their choice is arbitrary. We divide the star into a large number of concentric spherical layers, so that within the limits of each layer the density and temperature do not change very much. It is evident that knowing ρ and T, we can immediately find the gas pressure in each layer. Moreover, it is easy to find the acceleration due to the gravitation force and the weight of each layer.

We shall now assume that the star is in an equilibrium state. This means that the gas pressure in each layer must balance the weight of all higher layers. This condition can be easily checked by computation. If it is satisfied, then the chosen functions $\rho\,(r)$ and $T(r)$ correctly describe the model of the star. If this condition is not satisfied, it means that the choice of these functions is incorrect. We assume that the weight of the upper layers is larger than the density in certain layers of the star. We change the functions $\rho\,(r)$ and $T(r)$ so that the pressure will be higher in the layers where it was insufficient and we again check the equilibrium of the stellar model. If the gas pressure is still larger or smaller than the weight of the higher layers, we again change the functions $\rho(r)$ and $T(r)$. This process is called an iteration. We chose the functions $\rho(r)$ and $T(r)$ until the equilibrium condition is satisfied in all layers. If the choice of the initial values of these functions was good, only a few iterations are needed.

To calculate this by hand is a very tedious process, but for a computer it is a very simple problem. It is not complicated to write a program which automatically checks the equilibrium condition and also automatically changes the functions $\rho(r)$ and $T(r)$ so that the good values are rapidly found.

We mentioned above the condition of hydrostatic equilibrium: the necessity that the pressure in a given layer balances the weight of all higher layers. In stars one other equilibrium condition is necessary which must be taken into account in the calculation of a model: the equilibrium condition between the generation and transfer of energy.

We will start again from the beginning: we have the functions $\rho(r)$ and $T(r)$ and a stellar model divided into many layers. Since we know the density and pressure in each layer we can first calculate the amount of thermonuclear energy released in each layer (with formula 14) and second find the opacity of each layer. It is evident that to satisfy the equilibrium condition each separate layer must allow energy generated in all layers inferior to it to pass. We verify this using the computer in the same way as we have done to check the hydrostatic equilibrium. If the condition is satisfied, everything is alright. If not, we must again change the functions $\rho(r)$ and $T(r)$ so that we approach values which satisfy the condition and the thermal equilibrium, i.e. we use our iterative method here also.

In the calculation of stellar models the functions $\rho(r)$ and $T(r)$ are tested simultaneously for both equilibrium conditions and so with our iterative method we satisfy simultaneously the equilibrium conditions. Of course the volume of calculations increases, but for a computer with a large memory this does not represent a great difficulty.

Having obtained one stellar model as a result of several iterations, it is easy to obtain another model, for example, with a slightly different mass using as a first choice of $\rho(r)$ the value from the first model increased by a factor corresponding to the variation of mass. For the calculations of the second model we need less iterations than for the first.

The iterative method has one more important advantage: it allows us to follow the evolution of stars. We assume that a certain stellar model was calculated for a given chemical composition and that we have satisfied the equilibrium condition. After a certain time the hydrogen in the interior of the star burns up, the energy released decreases, and the equilibrium conditions are changed. One must calculate a new stellar model, taking into account the lack of hydrogen in its interior. This can be done easily with the iterative method taking the initial functions $\rho(r)$ and $T(r)$ and changing X and Y correspondingly to the amount of hydrogen consumed in each stellar layer within a given lapse of time. Thus, it is easy to build an evolutionary sequence of models of a given star.

Finally, the application of this method allows us to demonstrate some properties of the inner structure of stars. For example, while checking the equilibrium between the energy release in the inner stellar layers and its transfer to more external layers it appears that any choice of the functions ρ (r) and $T(r)$ will not allow us to satisfy the equilibrium condition. This could mean that energy transfer by radiation is not sufficient and that in these layers energy is transferred by convection. In this case the pressure and density of the gas in convective layers are given by formula (10) and must be taken into account in the calculation of stellar models. It can further appear that the pressure of a degenerate electron gas is greater than that calculated with Clapeyron's formula—this property must therefore be taken into account in the equilibrium calculations.

All this means that in the calculation of stellar models one must take into account many factors, but the physical notions of stellar model calculations stay as simple as those described above. One only needs a good powerful computer and also detailed tables of thermonuclear reactions and of opacities.

We have already mentioned that thousands of diverse stellar models have been calculated. Of course it is not necessary to describe here all these models. We shall give only general results.

We will first look at models of the main sequence stars. All these stars exist because hydrogen burns—is transformed into helium—in the central parts of the stars. The structures of main sequence stars appear to be simple, but they are different for stars of different masses. In the upper part of the main sequence, specifically in stars with masses larger than that of the Sun, there is a convective nucleus and the greater the mass of the star the greater also is the relative mass of the convective nucleus. In stars in the lower part of the main sequence (masses smaller than that of the Sun) there is no convective nucleus, but there is a convective zone close to the surface. Stars with very small masses (smaller than $0.3M_\odot$) are entirely convective. Figure 10 shows the variation of the relative mass of a convective nucleus and a convective zone depending on the mass of the whole star.

These properties can easily be explained. In massive stars the energy release occurs in the carbon cycle. In this case the power of the energy release depends very strongly on temperature ($\varepsilon \sim T^{20\cdots}$) and therefore almost all of the energy release is concentrated in the very centre of the star. Here a strong energy flux is generated which must pass through a small sphere surrounding the energy emitting region. Energy transfer by radiation is not sufficient and a convective nucleus appears. On the other hand, in stars of small mass, energy is released in the proton reactions and depends less on temperature ($\varepsilon \sim T^4$); therefore in this case the energy transfer by radiation is maintained in the centre. In the outer layers the temperature is low, the opacity of the matter high, and radiative transfer appears to be insufficient. A convective surface zone is formed. Such a convective zone also exists in our Sun.

84

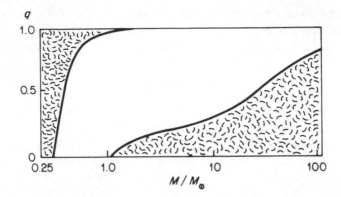

Figure 10 Distribution of convective zones and convective nuclei in stars with different masses. The value q indicates the quota of stellar mass occupied by the convective zone

The numerical values for the parameters of main sequence stars are given in Figure 11. In fact this graph represents the theoretical mass–luminosity relationship. The lower curve corresponds to models in which the hydrogen abundance is the same all over the star—this is the so-called initial main sequence. The upper curve describes stars in which hydrogen is almost entirely consumed in the central parts. The dashed curve represents stars of smaller mass, showing their positions after 10^{10} years of hydrogen consumption. During the process of hydrogen burning the mass of the star does not change, but the luminosity increases slightly (vertical arrows). For each given value of mass we indicate: the central density ρ_c in grams per cubic centimetre, the central temperature T_c *in millions of degrees, the stellar radius R* in units of the solar radius $R_\odot = 7 \times 10^{10}$ cm, the relative mass of the convective nucleus M_{conv}, and the time of hydrogen burning t_H in years. Also given is the transition from the carbon to the proton cycle ($\varepsilon_{pp} = \varepsilon_{CN}$), the place where the convective nucleus disappears and totally convective stars appear.

The reader can compare Figure 11 with Figure 3 and confirm that the theoretical mass–luminosity curve is in good agreement with that observed (the scales in Figure 11 are enlarged for convenience).

Figure 11 also shows a model of an 'initial Sun'. Its parameters are: $\rho_c = 90 \, \text{g cm}^{-3}$, $T_c = 13.9$ million degrees, $R = 0.87 R_\odot$. This model is calculated with the condition that hydrogen abundance is the same over the whole star and represents $X = 0.70$ (it has been also assumed that $Y = 0.28$, $Z = 0.02$). However, the actual Sun has already existed for at least 4.5 milliard years and therefore this model is not valid for the Sun with its present composition. A few models of the actual Sun were calculated with the assumption that hydrogen abundance is smaller in its centre than at its surface. In one of these models the following values were used: $X = 0.5$ in the centre, $\rho_c \approx 134 \, \text{g cm}^{-3}$, and $T_c \approx 14.6$ million degrees. The external convective zone occupies 15 per cent. of the radius but contains only 2 per cent. of the mass.

A very important fact is that while hydrogen is burning in the central parts of

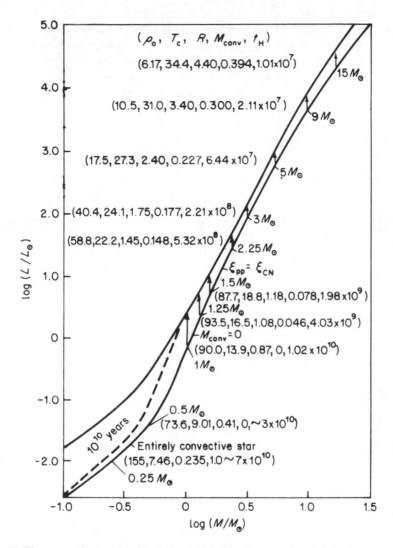

Figure 11 Theoretical mass–luminosity relationship. Lower curve—initial composition relative to hydrogen burning, upper curve—state at which hydrogen is almost entirely burnt in the convective nucleus. Dash line—position of stars with small mass after hydrogen burning for 10^{10} years. In brackets: ρ_c = central density, T_c = central temperature, R = stellar radius, M_{conv} = relative mass of the convective nucleus, $_H$ = time of hydrogen burning. The changes in thermonuclear reaction cycles $\varepsilon_{pp} = \varepsilon_{CN}$ are also marked, as well as the disappearance of the convective nucleus and appearance of entirely convective stars

stars their external parameters (luminosity and radius) only change slightly. This explains why there is such a distinct main sequence. The distance between the curves on Figure 11 increases with a decrease of mass, but if we consider the limited time of existence of these stars (about 10^{10} years) then the band is quite narrow. Of course, the calculations depend also on the initial chemical composition assumed. Figure 11 represents the main sequence stars of galactic clusters where we assumed $X \approx 0.7$ and $Z \approx 0.02$. Stellar models of subdwarfs belonging to globular clusters were also calculated. In this case the abundance of heavy elements is much smaller and there is more hydrogen One can even assume $Z = O$ and $X \approx 0.9$. Stellar models of subdwarfs with $Z = O$ differ from the models of Figure 11 by the fact that the convective zone decreases strongly and even disappears. The luminosity of subdwarfs with a given mass is smaller than that of ordinary stars, whereas the central temperature and density are approximately the same. However, calculations do not give definite results, mainly because there are no precise data on the initial abundance of helium.

We shall consider all of these models again in Chapter 10 when studying the evolution of stars. As the stellar models described above are notable for their comparatively simple structure, it is not possible to elaborate them in the same way as those for red giants or supergiants. It appears that the models of these stars have a complex structure and that this structure is a consequence of stellar evolution.

As an example we give here the parameters used for one model of a giant star; the mass is equal to 1.3 times the solar mass, the luminosity is 226 times greater than the luminosity of the Sun, and the radius is 21 times greater than that of the Sun. This model serves only as an example. The description given below shows in which conditions it was possible to obtain a model satisfying the equilibrium conditions as well as giving the values for the fundamental parameters.

In the centre of the star is an isothermic nucleus with a constant temperature of 40 million degrees. In this nucleus there is no hydrogen; therefore no thermonuclear reactions occur and no energy is released. It is evident that all the hydrogen has already been 'burnt' in the nucleus and is composed almost entirely of helium with small amounts of heavy elements. An isothermic nucleus represents 26 per cent. of the whole stellar radius. The density is therefore very large—the central density is 3.5×10^5 g/cm^3. Consequently the central isothermic nucleus of the star is composed of degenerate electron gas; in other words, one can say that in the centre of a red giant there is a typical white dwarf.

The degenerate isothermic nucleus of a giant is surrounded by a thin layer— about 0.08 per cent. of the stellar radius—in which energy is released by ordinary thermonuclear reactions. Inside this thin layer the temperature falls very sharply from 40 to 25 million degrees and the density changes from 70 to 14 g/cm^3. The energy releasing layer is surrounded by another layer occupying 8 per cent. of the stellar radius and containing 5 per cent. of the total stellar mass in which energy is transformed by radiation. Finally, the other 92 per cent. of the stellar radius and 70 per cent. of the stellar mass represents a large 'inflated' envelope in which energy is transferred by convection and where the temperature does not reach even one million degrees. The formation of a convective zone is due to the same reason as for the main sequence red dwarfs:

the low temperature results in the stellar matter being opaque and this 'locks in' the radiation. Thus the characteristic nature of red giants is, to a high degree, an inhomogeneous structure with, on the one hand, a very dense nucleus and, on the other hand, a very extended envelope. For this reason formula (5) cannot give a correct value for the central temperature of these stars—in the deduction of this formula we assumed a more or less homogeneous stellar structure. We must also consider the degenerate stellar matter in the central parts of giant stars.

The model of a giant star is not entirely satisfactory—the complex structure of these stars introduces an important factor of arbitrariness into the calculations. It has not yet been possible to calculate a model of a supergiant. The calculations are particularly difficult for the transition zone and the energy releasing layer. We shall study the models of giant stars in Chapter 10.

8

Variable and non-stationary stars

In the preceding chapters we have studied the inner structure of stationary stars. In the true meaning of the word, stationary stars, i.e. stars that do not change their composition, generally do not exist—because every star evolves. However, if the changes in stellar composition (e.g. the transformation of hydrogen into helium) occur very slowly, during hundreds of millions or milliards of years, then these stars can be called stationary stars. The majority of stars in space change much more rapidly than that. First, there are variable stars which are different because their light (i.e. luminosity), radius, and spectrum undergo more or less regular variations. Second, there are 'explosive' stars which flare and throw off gaseous envelopes from time to time. There are also many different types of stars which show variations in the spectrum—sometimes rapid, lasting a few minutes, sometimes slower—which are often not understood. It is difficult to classify these stars as we still do not know the basic physical processes which bring about these sharp variations. For this reason they are usually classified only according to their external indications. We are interested here in stellar physics and therefore limit ourselves to data which can be explained from the physical point of view.

All 'rapidly changing' stars (on the average one star in 150 000 undergoes rapid changes) can be divided into two groups: periodic variables and properly non-stationary stars, with which we also associate explosive variables. The study of variables and non-stationary stars is interesting for a number of reasons. It is evident that in the process of slow stellar evolution an accumulation of numerous changes leads to qualitative changes—evolution gives way to 'revolution', stars in such a 'revolutionary' state also being non-stationary stars. The violent changes in a non-stationary star that we observe allow us to penetrate deeper into the physical processes which take place in it, as well as to check the validity of our theory for the inner structure of stationary stars. In certain types of non-stationary stars cosmic rays and heavy chemical elements are apparently generated.

Periodic variables are the 'lighthouses' of the Universe: they allow us to find, with the help of the period–luminosity relationship given below, the distance to

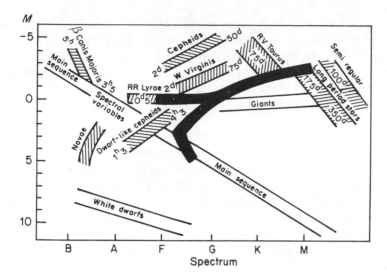

Figure 12 Spectrum–luminosity diagram for variable stars. The white bands represent stationary stars of type I population (i.e. relatively young); the black bands represent stationary stars (relatively old) of type II population. The bands with horizontal hatching are variable stars of type I population; the bands with vertical hatching are old variables and non-stationary stars of type II population. Periods of variable stars are shown (in hours or days). The explanation of the types of variables is given in the text

far away stellar systems. It is not surprising that astronomers pay particular attention to variable and non-stationary stars but very few of the observed particularities of these stars have been explained. At the present time only the theory of pulsations of periodic variable stars has been developed fairly well. We shall briefly explain it.

Periodic variables, as their name indicates, change their luminosity with a more or less regular period. Figure 12 shows a spectrum–luminosity diagram for all variable stars given by the well-known American astrophysicist O. Struve. Each group of variable stars is marked by a band, indicating the dependence of the luminosity on the surface temperature (or spectrum). The mean periods of oscillation (in days) are also shown, as well as the limits in which the periods of a given type of variable star are situated if the oscillations of different variable stars or the same type have different periods.

First of all, wc note that the positions of variable stars and stationary stars do not coincide. This means that their structures are different, and since there is a comparatively small number of variable stars we can say either that the phenomenon of variability is linked to the brief change in the very structure of an ordinary star or that these are particular stars. In the left-hand part of Figure 12 there is a rather important interval of possible oscillation periods which are small. The hotter the star, the shorter its period. In cool stars the periods are large. It appears also that if in hot stars the oscillations are comparatively regular in cool stars they are irregular.

All the types of variable stars given in Figure 12 have diverse spectral characteristics. We have already seen that many phenomena are not understood. For this reason we will proceed in the following way. We shall describe in detail the aspects of variables for which there are explanations for the phenomenon that occurs and expose the basis of the oscillation theory for these stars. Then we shall briefly describe the particularities of the 'unexplained' variable stars.

The most 'comprehensible' and perhaps therefore also the most interesting type appears to be the class of variable stars which is designated by the name 'cepheids' (from the constellation Cepheus where a variable star of such a type was discovered for the first time). This class is composed of different stellar groups. In Figure 12, in the cepheid class (sometimes for distinction they are called 'classical cepheids') are W Cep and RR Lyrae stars and each group of stars is again divided into subgroups.

All stars of the cepheid type have a constant period. A particular stability distinguishes a comparatively small subgroup of stars belonging to the type I population, the RR Lyrae stars. The oscillation period of these stars remains rigorously constant during tens of millions of pulsations. Of course, the majority of the RR Lyrae stars, namely those belonging to stars of the type II population, is not very stable: after a few years they can change the period or phase of oscillation. Considering that the oscillation period of these stars is on average 12 hours, we find that their period does not change in approximately one thousand pulsations. This is not very bad; in other variables the periods and phases can change much more rapidly. Besides, in cepheids the variations of period are small and the oscillations can almost always be considered sufficiently constant.

In classical cepheids the oscillation periods coincide with an interval from 2 to 10 days but stars with a period of 8 days (a mean period) are the most frequent. Figure 12 shows that the period increases with an increase in luminosity and a decrease in surface temperature of the star. This is a very important property of cepheids which has not yet been entirely explained. Studying the observational data, astronomers have found a statistical relationship between the period of a variable star P (expressed in days) and the logarithm of its luminosity (see Figure 13). This relation allows us to consider the cepheids as 'lighthouses of the Universe'. In fact, having determined the period by observations, an astronomer can find the luminosity of the star from Figure 13, and knowing its luminosity he can easily determine the distance. We must emphasize that Figure 13 is not a theoretical result but an 'experimental' one. Moreover, one must keep in mind that Figure 13 is valid only for classical cepheids. For Virga stars a similar relationship exists (this is clear from Figure 12). These stars, with identical periods, are on the average one and a half stellar magnitudes smaller than the classical cepheids. All RR Lyrae stars have the same luminosity, which is forty times greater than that of the Sun. W Virginis stars differ from classical cepheids in being less bright and also having larger mean periods—about 15 days. We have already mentioned that these stars belong to the type II population.

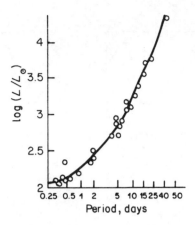

Figure 13 Relationship between the cepheid luminosity and the period of pulsation

All three groups of cepheids, and in general also their subgroups, are notable for the variation of the stellar brightness (Figure 14), the characteristic change of their surface temperature, and finally the change of velocity of the stellar surface motion. The latter is measured from the radial velocities of atomic motions in the upper layers of the star on the basis of the well-known Doppler effect (Figure 15). This point will be considered in more detail.

The reader certainly knows the Doppler law: the frequency of light emitted by a moving source changes in proportion to the velocity of the source. Using this law, we can determine from observed stellar spectra the velocities of the atoms which are at the surface of the star. In fact the atoms emit spectral lines of entirely determined frequences. Having measured the frequency of the same atom on Earth we can immediately find from Doppler's law the velocity we are interested in. In stationary stars the velocity of an atom at the surface is evidently on average (if we exclude thermal motion) equal to the velocity of the star.

When, in a similar way, the velocities of variable stars were measured it appeared that these velocities changed within one period. The results have shown that the stellar surface changes its position with respect to the centre of the star, which of course moves with constant velocity, within one period. In other words, variable stars pulsate—they expand and contract. This explains the change observed in the surface temperature of the star; its position and luminosity change which, according to formula (1), brings about oscillations in the value T_e. Figure 15 gives the variations in the visible surface velocity for cepheids. The plus sign corresponds to a greater distance of the surface from the observer (the contraction of the star) and the minus sign corresponds to a closer distance (the expansion of the star). In the graphs of Figure 15 the zero is not always situated in the middle because here the stellar motion was not considered as a whole. If the curve is placed so that the zero is precisely in the centre the velocity of motion of the whole star and the velocity of expansion

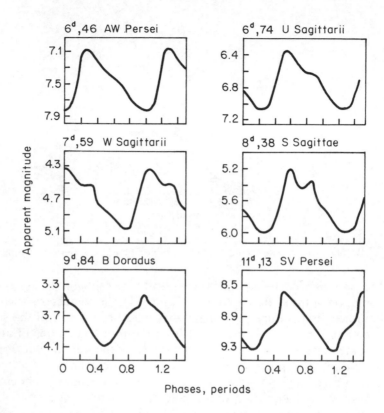

Figure 14 Brightness curves of a few cepheids. The visible stellar magnitude is given relative to the phase that is the portion of the period calculated from a certain fixed point. The periods are shown in days

and contraction of its envelope can also be determined. It appears to be about 10 km/s. Note that the variations of velocity as well as of brightness curves are, in general, not symmetric–we cannot compare the pulsations of a star to the regular oscillations of a gas sphere. These particularities of the brightness curves are explained by the behaviour of the very exterior stellar layers (which we consider here) and therefore if we do not consider the particularities of the velocity curves we can come to the conclusion that a variable star is a gas sphere pulsating with a determined period.

The pulsation theory for stars is quite complex but we have already seen many times that the basic relations can be obtained in an elementary although not very rigorous manner.

Let us consider a concrete problem: we will find the oscillation period of a gas sphere with radius R and mass M pulsating under the action of the gravitational force. To solve this problem we can use the analogy between pulsations of gas spheres and oscillations of a pendulum.

First of all we recall the well-known rule for oscillations of a mathematical pendulum found by Huygens. The oscillation period is directly proportional to

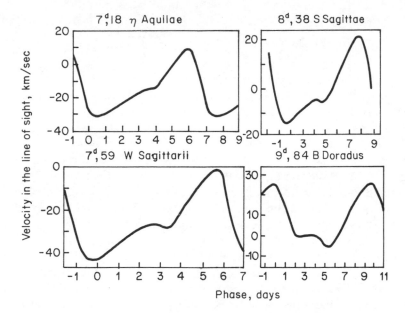

Figure 15 Velocity curves of some cepheid surfaces. Observed velocities are given relative to the oscillation phase which is given here in days. Note that the brightness curves and those of velocities do not coincide for the same stars

the square root of the value of the acceleration due to gravity, that is

$$(44) \qquad P = 2\pi\sqrt{\left(\frac{l}{g}\right)}.$$

Here P stands for the oscillation period, l for the length of the pendulum, and g for the acceleration due to gravity. We must note that this formula is quite universal. It can also be applied in other problems to find the oscillation period, when one need only choose a value equivalent to the length of the pendulum l.

Of course the oscillations of a gas sphere are not very similar to the oscillations of a pendulum, but taking into account the universal character of the Huygens formula we can try to use it to find the oscillation period of stars. The 'length of the pendulum' will be taken to be equal to the stellar radius; the acceleration due to gravity in the star is, as we already know, equal to $g = fM/R^2$. Introducing these values into the Huygens formula we obtain a formula for the oscillation period of a 'pendulum-star':

$$(45) \qquad P = 2\pi\sqrt{\left(\frac{R^3}{fM}\right)}.$$

This formula can be simplified if we remember that the mean density of a star is $\bar{\rho} = 3M/4\pi R^3$. Then we obtain

$$(46) \qquad p\sqrt{\bar{\rho}} = 2\pi\sqrt{\left(\frac{3}{4\pi f}\right)} = \sqrt{\left(\frac{3\pi}{f}\right)}$$

Thus for oscillations of a gas sphere in a gravitational field the product of the oscillation period over the square root of its mean density is a constant. This is a very important formula. As a matter of fact, in the case of oscillations of gas spheres it changes the law of Huygens. We shall now see whether this formula can be applied to oscillations of cepheids. For this we should compare the products of an observed period with the square root of the mean density, obtained for very different pulsating stars. Such comparisons have been done many times and it appears that if one takes, for example, classical cepheids for which the periods are within an interval of 2 to 35 days and the densities of different stars differing by three orders of magnitude then for them the product from (46) does not change by more than one and a half times. It is true that the numerical value of the constant is a little smaller. With formula (46) we obtain $\sqrt{(3\,\pi/f)} = 0.12$ (if the period is expressed in days and the mean density in grams per cubic centimetre), whereas from the observations this value is between 0.4 and 0.6. This can be explained by the fact that in deducing (46) we overestimated the 'length of the pendulum', considering it to be equal to the stellar radius. In fact, as only the surface layers are pulsating, this leads to a decrease of the constant in formula (46).

Formula (46) is very important for the theory of stellar pulsations and therefore astronomers have tried to make it more precise by taking into account the thickness of the pulsating layers and the density distribution in the stellar interior. As a result they obtained the relationship

$$(47) \qquad P\sqrt{\rho} = 0.026 \left(\frac{R}{R_\odot}\right)^{1/4} \left(\frac{M_\odot}{M}\right)^{1/4} ,$$

from which one can immediately obtain a formula determining the oscillation period of the star relative to the acceleration due to gravity on the stellar surface and its mass:

$$(48) \qquad P = 0.022 \left(\frac{g_\odot}{g}\right)^{7/8} \left(\frac{M}{M_\odot}\right)^{1/8} \qquad \text{days.}$$

The oscillation period for cepheids is determined mainly by the acceleration due to gravity on their surface—this verifies the fact that it is mainly the upper layers of the star which pulsate. It is important that formula (46) is approximately correct for any pulsating star. The more precise theory of (47) and subsequently also of (48) has only been formulated for cepheids.

Our Sun does not pulsate—its structure is different from the structure of cepheids. If it did start to oscillate, however, the pulsation period would be one and a quarter hours (according to formula 46) or 32 minutes (according to formula 48), i.e. shorter than for the shortest period of an RR Lyrae star. We emphasize again that pulsating stars differ in their structure from ordinary stationary stars although it is possible that this difference is not very large.

The formula for the oscillation period was easy to obtain and this formula appears to be in good agreement with the observational data, but the remaining peculiarity of pulsations in variables is very difficult to explain. For example, we see from the velocity of motion of stellar surfaces that the amplitudes of the pulsations are small and from mechanics we know that any small oscillation must be of a sinusoidal nature. However, we have already seen that the brightness curve does not resemble a sinusoid. Another peculiarity is perhaps even more important. One could expect that the star reaches maximum brightness either at the moment of the maximum contraction, when the stellar matter is hotter than at the moment of maximum expansion, or at the moment of expansion, when its surface is at a maximum. Observations of temperature variations on stellar surfaces show that these variations are more important than variations in the radius; therefore we can expect a maximum brightness at the moment of maximum contraction. In fact, it appears that the star reaches maximum brightness at the moment when its surface is closest to us with the highest velocity, i.e. somewhere in the middle between the maximum contraction and expansion. This peculiarity is often explained by the fact that oscillations close to the stellar surface are different from oscillations in the interior. What happens to oscillations of a star near its surface resembles in a way the phenomenon of surf: a wave far away from the shore resembles a sinusoid; coming closer it changes its form and the different parts of the wave profile move in a different way—the crest catches up with the base and the wave becomes very short. Below we shall give details of this phenomenon based on results of numerical calculations of stellar pulsations.

Of course the most important problem of pulsating stars is to find the causes of pulsation, in other words, to answer the question: why do certain stars pulsate and others not? Although we cannot give a complete and detailed explanation, the causes of pulsations are known and astronomers can calculate them. Let us come back to the pendulum. If we push it, it will oscillate but it soon comes to a stop—the friction in the bearing of the pendulum and the resistance of air will take away all its energy. The pendulum will oscillate for a long time only if the energy is supplied by shocks or continuously, thereby compensating its loss by resistance. If we push the pendulum each time at the moment when it is at the greatest distance we support its oscillations with a 'periodic' force. However, a pendulum can oscillate for hours without stopping, under the action of a constant force, e.g. the tension of the spring in an ordinary clock with weights for a pendulum. Such oscillations due to a periodical force are called auto-oscillations.

A star may be compared with a pendulum from this point of view. Part of the energy of motion is converted into heat and cannot be reconverted to kinetic energy (the same change that is a result of the friction of the pendulum). If we 'contract' the star and then 'release' it, it will start to oscillate (pulsate), but the oscillation amplitude will rapidly decrease and after only five to ten expansions and contractions it will return to the initial state. Pulsating stars vibrate much longer; consequently there should be some kind of force compensating for the energy loss and maintaining the oscillations at a certain level. It is difficult to imagine the force which, for example, at the moment of maximum expansion would push the surface of the star—there is no periodical force in pulsating stars. However, there is no need to look for a constant force acting continuously—the force is the result of energy release in thermonuclear reactions and the passage of this energy through the entire thickness of the star. Thus we can assume that a pulsating star is an auto-oscillating system, the same as, for example, in a clock.

To complete our analogy we must find the mechanism which acts in the same way as the counterweight in the clock and leads to oscillations, 'portions' of energy with periods corresponding to the oscillation period. In this case two possibilities exist. We know the amplitude of oscillations at the stellar surface but we do not know the beahviour of oscillations inside the star. We first assume that in the central parts of a star the oscillation amplitude is at least not very small. Then we obtain the following. As the stellar layers contract the temperature will increase and according to data from Chapter 4 the release of thermonuclear energy increases strongly due to the strong dependence of this release on temperature. Now the energy release increases the gas pressure in the layer and the contracted layers tend to expand with an even greater force than when they were contracted. This is the first mechanism maintaining and even pushing the stellar oscillations. Of course the central parts of the star must be pushed with a certain amplitude, which is not easily done as they are densely 'packed'. The calculation effectively shows that in energy releasing layers the amplitude of oscillations is practically zero.

The second mechanism of oscillation, sometimes called the 'valve mechanism', is somewhat more complex, but more effective, simply for the reason that it is linked to oscillation of stellar layers close to the surface. In this mechanism oscillations are maintained even if they affect about 1 per cent. of the stellar mass and about ten per cent. of its radius. To explain this mechanism we must start almost from the beginning.

In Chapter 3 we saw that the opacity of stellar matter depends on density and temperature and we described this relationship using Kramer's formula: $\varkappa \sim \rho/T^{7/2}$. It is true that in the outer layers of the star the matter behaves in a more complex way and there is a different relationship between $\varkappa$ and ρ and T (see Chapter 3). We shall presume that the stellar matter is such that Kramer's formula is valid, and this allows us to explain the nature of the 'valve' mechanism.

Let us see what happens to the opacity of a certain layer inside a pulsating star when it contracts and expands. It is evident that we must know how the temperature and density change in this layer. We first assume that when contraction and expansion takes place the thermal energy is maintained—in this case one talks about adiabatic pulsations. We already know that for adiabatic variations the gas pressure is proportional to the density of degree γ

(see Chapter 3). Since $\rho = (A/\mu) \rho T$, the temperature is proportional to the density of degree $(\gamma - 1)$ (that is $T \sim \rho^{\gamma-1}$). Now, as in formula (10), we can write an analogous relationship between the variation of the opacity proportional to the contraction and expansion of the gas layer:

$$(49) \qquad \frac{\varkappa_1}{\varkappa_2} = \frac{\rho_1}{\rho}\left(\frac{T_2}{T_1}\right)^{7/2} = \left(\frac{\rho_1}{\rho_2}\right)^{(9-7\gamma)/2}$$

This relation yields a very important conclusion: if the exponent of the adiabatic line of the gas $\gamma > 9/7$ then on contraction of the gas its opacity decreases. At first sight this seems to be strange but it is easily understood. By contracting a gas we increase its temperature and consequently strip electrons off the nuclei—the gas becomes more transparent. However if $\gamma < 9/7$ then on contraction of the gas its opacity increases—the temperature increase is insignificant and the higher density of the gas increases the opacity. This condition ($\gamma < 9/7$) seems very strict and at first sight it is encountered nowhere, as in an ordinary single atom of gas $\gamma = 5/3$. However, we assume that somewhere in stellar interiors there is a layer with $\gamma < 9/7$. What brings this about?

We contract a star with a not too thin layer composed of a gas with $\gamma < 9/7$. Then, except in this layer where the opacity increases, the opacity of the whole star decreases. As a result the layer will retain the energy coming from the central parts of the star and absorb it. This will bring about a temperature increase in the given layer and compel it to expand and push the layer of ordinary gas situated above with a greater force than was used to contract the layer of gas. Thus the layer with $\gamma < 9/7$ acts like a 'valve', withholding thermonuclear energy radiation at the moment of contraction and compelling part of this energy to be used to support the oscillations. According to the example given by S. A. Zhevakin (who formulated this model for the theory of pulsations) the action of a layer with $\gamma = 9/7$ resembles the work of a diesel motor where the fuel (an addition of energy) is also injected at the moment of contraction. However, here the role of the fuel is played by the absorbed radiation coming from the interior of the star. We can also establish an analogy between the action of a layer with $\gamma < 9/7$ in a star and the action of a balance-wheel in a clock.

In order to maintain pulsations in a star it is therefore necessary to have a gas layer with $\gamma < 9/7$. Can such a layer be found in real stars? On the surface of stars such as the Sun, there is single atom gas with $\gamma = 5/3$. On the surface of cool stars there are molecules, but even if we presume that all atoms are united into molecules of two atoms then even here $\gamma = 7/5$. All the values for γ are greater than $9/7$. In the stellar interiors, gas is ionized—divided into electrons and atomic residues—but behaves like single atom gas. Consequently here also $\gamma = 5/3$. Thus almost in the whole star $\gamma > 9/7$. However, it appears that a layer with $\gamma < 9/7$ also exists in stars.

It is evident that somewhere at not too great a depth under the stellar surface there should be layers where the transition from weakly ionized gas at the stellar surface to entirely ionized plasma in its interior takes place. For example, in regions with a gas temperature of about 10–20 thousand degrees, intense ionization of hydrogen occurs. At higher temperatures hydrogen is almost not ionized; at lower temperature it is almost entirely ionized. At a temperature of about 40 thousand degrees helium is strongly ionized (its second electron is torn off). The regions in stellar interiors with similar temperatures are called hydrogen ionization zones and helium ionization zones. It appears that in such zones of ionization the parameter can be much smaller and that the condition $\gamma < 9/7$ is possible there.

It is easy to imagine that γ will be close to unity if on contraction of the gas the temperature increase is small (since here $T \sim \rho^{\gamma-1}$). This will also be true in the ionization zones. We will now contract such a zone. On contraction the temperature should increase and consequently a greater facility to ionize atoms is acquired. The energy used is that gained on contraction. In other words, an important decrease of γ occurs on contraction of a gas 'on the threshold of ionization' when only a small increase in the energy of the particles is sufficient to increase strongly the ionization.

Thus we see that, at least in principle, the hydrogen and helium ionization zones can serve as a valve mechanism withholding the flux of thermo-nuclear energy coming from the stellar interior on contraction of these zones. The fact that these zones are situated near the stellar surface makes them effective for the maintainance of the oscillations in real stars. We cannot give numerical estimations here of the mechanism considered, partly because we have examined only the principle aspect of the phenomenon—in practical calculations one must also take into account the fact that Kramer's opacity law is not sufficiently precise—and partly because the calculation for the ionization zones is quite complicated.

Our qualitative estimation is as follows. Although there is much hydrogen in stars, the zone of helium ionization is more effective. The hydrogen zone is too close to the surface and the opacity coefficient is smaller there. In a number of cases it does play a role, but the fundamental part is played by the zone of second ionization of helium (when the second electron of the helium atom is torn off). A helium abundance of 15–30 per cent. is sufficient in a star and the ionization zone can serve as a valve for the maintaining pulsations. In order to obtain the correct value of γ to permit the action of the 'valve mechanism' (that is $\gamma < 9/7$ in the case of Kramer's formula) a few more limits concerning the gas composition in the ionization zones are needed. This, in turn, leads to limits imposed on the acceleration due to gravity in the surface layers. From this results the known observed fact that only certain stars of a particular composition pulsate.

One problem of the theory of pulsating stars which we have not yet considered is the problem of determination of the oscillation amplitude. On the basis of considerations given above we can also study this problem, although it is more difficult. In fact, we know that everywhere inside the ionization zone the pulsations are linked to the transformation of kinetic energy into thermal energy—in other words, a dissipation of energy occurs. In the ionization zones the situation is different; here the energy of the radiation flux is converted into kinetic energy—we say that 'negative dissipation' occurs. Positive and negative dissipations depend in different ways on the amplitude of oscillations. In a cepheid oscillating with a constant amplitude the negative dissipation should entirely compensate (but not exceed) the positive dissipation. It is clear that this can only occur at a determined oscillation amplitude.

This phenomenon can also be explained in the following manner. Imagine that in a star in which pulsations can exist there were initially none. What incidental causes brought about the very small contraction of the ionization zones? At first the 'valve mechanism' scarcely 'pulsated' the energy. Then oscillations with very small amplitude started. Such oscillations cause very little energy losses. The 'valve' mechanism maintained the pulsations of oscillations. Gradually their amplitude increased until finally a state existed in which the positive dissipation was so great that it 'ate up' all the energy released in the ionization zones. From that moment on, the star will oscillate with a constant amplitude until changes in the stellar structure upset this equilibrium. This process of the establishment of the oscillation amplitude can only be studied by computation. A few such computations have been done recently. We shall give the results of two of them.

The first concerns a star with a mass of 6.75 solar masses, a radius 48 times greater than the solar radius, and a luminosity equal to $2200L_\odot$. The amplitude of established oscillations of the radius is 2 per cent., the oscillation amplitudes of the surface temperature is 4.5 per cent., and the luminosity is 7 per cent. In the helium ionization zone, playing here the basic role, the oscillations are sinusoidal; in the upper layer of the star an asymmetry appears.

The second concerns a star with parameters $M = 0.4M_\odot$, $R = 4.9R_\odot$, and $L = 390L_\odot$. In the computation we presumed that the initial oscillation amplitude of the radius equals 10 per cent. After forty oscillations the amplitude grew to 18 per cent. The luminosity oscillation amplitude grew from 8 to 13 per cent. In this star two-thirds of the negative dissipation is generated in the zone of secondary helium ionization (at $T = 40$ thousand degrees) and the remaining third of the energy in the hydrogen ionization zone and the zone of first helium ionization ($T = 22$ thousand degrees). Both zones release within one oscillation period 7.2 per cent. of the energy generated by the whole star during the same time. From this amount 78 per cent. serve to compensate energy losses in oscillations of deeper stellar layers since the coefficient efficiency of a pulsating star, as that of an engine, is $7.2 \times 0.22 = 1.6$ per cent. This is not very much.

Figures 16 and 17 show computations by R. Christie of luminosity variations and of the velocities of stellar layers at different depths. The layers are numbered from a certain depth to the surface and are denoted by a number on the axis. The first layer is the helium ionization zone, layers 20 to 30 are inside the ionization zone, and layer 40 corresponds to the surface. We must note that the luminosity oscillations reach a maximum amplitude inside the ionization layers of hydrogen and helium where, according to what has already been said,

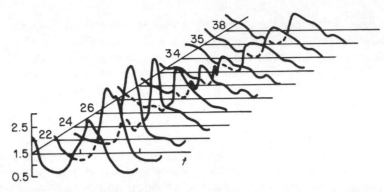

Figure 16 Energy flux (luminosity) variation curves for one period at different depths of the pulsating star. The unit of luminosity on the graph is 10^{36} erg/s. The luminosity oscillations are greater inside the star and smaller on its surface. Compare the graphs of Figure 16 and 14

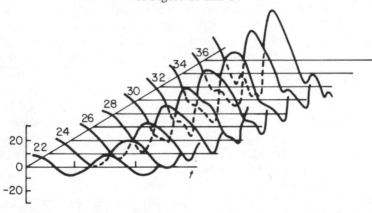

Figure 17 Variation within one period for the velocity of motion of the layers inside a pulsating star. The oscillation amplitude, given in kilometres per second, increases towards the surface. Compare with Figure 15

oscillations are also considered. On the other hand, the variation amplitude for the velocity of the gas is much greater at the surface. This phenomenon is reminiscent of surf where the height of the wave in the sea increases as it rolls towards the shore. In the same way the velocities of waves in stars increase as they 'roll' to the stellar surface. One can also note that the approach to the surface distorts the wave. At great depths the wave is sinusoidal; close to the surface the wave has a shorter rise time and longer fall-off.

We have paid a great deal of attention to the description of pulsating cepheids. This was done on purpose, since among all non-stationary stars they are the only ones for which we can say with certainty that we understand the processes occurring there. Of course there are pulsating stars for which we can only say that there are some oscillations but how and why they occur is not known. For example, in Figure 12 is a line representing RV Taurus stars. These are also pulsating stars with a period of about 75 days, but besides this period they also have a larger oscillation period of some thousands of days. The amplitudes of both oscillations change with time rather strongly. The brightness curve also has a complex and changing character.

Even more irregularities appear in observations of long period variables to which belongs, in particular, the well-known Mira Ceti star with a period of 406.95 days. The study of so-called semi-regular or irregular pulsating stars is very complicated. In each such star a few oscillations with changing amplitudes can occur and there are no maintained periods. The visible oscillations of the brightness of these stars are very big but this is linked to the particularities of their atmospheres. All these stars are red giants and supergiants in whose atmospheres there are many molecules. Within one period the percentage of molecules changes strongly, which leads to variations in the visible brightness. These stars are also characterized by emission lines, appearing from time to time, released by gas clouds rejected from their interior.

It is possible that the quality of pulsations of these stars has something in common with the pulsations of cepheids, but the complexity of the processes occurring in such stars has not yet permitted us to understand them and establish a numerical method.

The diagram of Figure 12 also shows pulsating hot β Canis Majoris stars which are characterized by a small oscillation amplitude and also the dwarf-like cepheids. For these stars no theory has been elaborated yet. One can hope that in the near future we will know more about these stars.

Even less concrete facts are known about another big class of non-stationary stars—the so-called flare stars. The general characteristic feature of these stars is that from time to time there is an outburst in such a way that their upper parts are blown off and the brightness grows rapidly. Then the expanded envelope is torn off the star and flies off into interstellar space. From the surface of the remaining part of the star a flow of matter continues but this gradually stops and the star returns to a stationary state. Stars of this type are also varied; their basic distinctions are in the scale and the nature of the flare. To a certain degree the Sun, and probably also all other stars, belong 'in miniature' to stars of such a type. It is known that on the Sun, particularly at the period of maxmimum solar activity, so-called chromospheric flares accompanied by rejection of matter occur from time to time.

Such flares probably exist in many stars, but cannot be observed against the background of ordinary, stationary radiation of these stars. There are also stars (called 'flare' stars, in particular of the UV Ceti type) in which flares are so strong that within 10 seconds their total brightness increases by ten to a hundred times. Then, after 10–20 minutes the star 'calms down' and returns to its initial state. In chromospheric flares on the Sun radioemission is generated—as in UV Ceti stars. However, the scale of all these phenomena is of course much greater in these stars than in the Sun. Another difference also exists. UV Ceti stars are red dwarfs of class M and are part of binary systems, whereas the Sun is a single yellow star.

Apparently, at the moment of a flare, a gas cloud is ejected from the flare star but the whole stellar surface is not affected. If the scale of the phenomenon is more important and if it occurs at a greater depth it cannot be called a flare but an explosion. In this case a great diversity of phenomena is observed, determined in the first place by the scale of the explosion.

There are stars in which explosions are comparatively small and occur in the surface layers—these are called nova-like stars. Some stars have large-scale flares where the explosion affects deeper layers of stellar interiors (a few per cent. of the radius). These stars are called novae. Finally, if the explosion affects an important part of the star we are dealing with a so-called 'supernova'. The name 'nova' is not quite appropriate and can be justified only from a historical point of view. Novae have, up to the flare, a very insignificant visible brightness. Such stars cannot be seen through any telescope. At the moment of the flare the brightness of a nova greatly increases and it becomes visible; it seems that this star appears at a place where it did not exist before. For this reason a nova is in fact not literally a new star but an old star in its final evolutionary state.

A numerical theory of nova flares has not yet been established. Many different hypotheses have been studied but astronomers still cannot reach a common opinion on the structure of flare stars and the reason for these flares. This is mainly due to the fact that we have very little observational data on nova stars. Presently in the Galaxy many hundreds of novae flare but we observe only one or two a year, and at times we do not even observe that number since the majority of novae flare a great distance from the Sun and we do not notice them. Usually novae are discovered accidentally. However, even if we do see a nova we cannot find out much about the star itself. It has never been studied and nobody even knew that this star would outburst. Nevertheless, a few things are known.

In a typical nova the brightness increases during the explosion by about ten thousand times. In nova-like stars the amplitude of the flare is much smaller— the brightness increases by only several tens of times. In supernovae, however, the amplitude of flares is indeed enormous—within a few days, and sometimes hours, one supernova radiates as much energy as the whole Galaxy composed of hundreds of milliards of stars. We do not know their luminosity before the flare and therefore the amplitude of brightness variation is not yet known. It has been discovered that stellar flares can reoccur and the smaller the amplitude of the flares the more often they occur. In nova-like stars flares reiterate during several tens and hundreds of days, while in typical novae this period should be hundreds and thousands of years (unfortunately this last conclusion cannot be checked as modern astronomy is still too young). Finally, supernovae probably flare only once in their lifetime.

Observations of the spectrum of a nova during the growth of its brightness show a blowing-off of the stellar surface layers at velocities of several tens and hundreds (up to a thousand) of kilometers per second. Probably a rather dense opaque envelope is detached from the star and ejected by the force of the explosion. The increase of brightness is due to the immense increase of the dimension of the stellar envelope.

When the blown-off envelope of the variable star reaches its maximum it starts to fall back into the star. However, this is not the situation in the case of a nova. After the maximum brightness of the nova the envelope continues to recede from the star. At the moment of maximum brightness the envelope is transparent. Its radiation therefore decreases. The brightness of the nova should diminish even further since the envelope of the star stops emitting (it consumes its supply of thermal energy) and the star, seen through the envelope, comes gradually to a state of rest. After the explosion of the main envelope, stellar matter continues to be rejected, often at a velocity greater than that of the fundamental, primary envelope. Nevertheless, in spite of the great velocities these rejections are already less intense and gradually the ejection of matter from the surface of the nova comes to a stop. The progressive acceleration of the star continues for several years and even several tens of years. We must note a propos that two types of nova exist: so-called rapid novae in which all changes occur in accelerated speed and slow novae with an inhibited development of the whole cycle. In general, the different novae show a great variety of details in their evolution.

The stellar matter ejected by a nova forms a nebula. The study of these nebulae is very interesting but goes beyond the limits of this book. In particular, by applying to this nebula the corresponding methods developed in the physics of nebulae and interstellar gas, the masses of ejected envelopes can be determined. It appears that at the first explosion a mass of about 10^{-3}–10^{-4} of the solar mass is ejected. This fact shows that in the process of explosion only surface layers of the star take part. In the following ejection of matter into interstellar space a mass of similar magnitude 'flies off'. The explosive energy of a nova can be approximately estimated by multiplying the mass of the envelope by half of the square of the ejection velocity (we can take a velocity of 1000 km/s). We obtain for the explosive energy the value of $\sim 10^{45}$ erg.

The causes of explosions in novae and similar stars are not as yet known. Many hypotheses exist. For a time, nuclear explosions were a very popular assumption. Let us assume that somewhere, at a determined depth inside the star, thermonuclear reactions take place. The velocity of the reactions depends strongly on temperature. On the other hand, the energy released in thermonuclear reactions is carried off by radiation. The velocity of energy evacuation is determined by the opacity which also depends on the temperature, but to a much smaller degree. If in the place where thermonuclear reactions occur the temperature inceases suddenly, though very little, then the energy release will sharply increase and the radiation will no longer be able to cope with its transfer because of the slight change in opacity. The temperature will increase even more, and also the pressure, which will induce an explosion. The energy of the explosion can be transmitted to the stellar surface by a shock wave. It is possible that the explosions of novae are linked to the transition from one thermonuclear reaction to another or, in general, to the reconstruction of certain stellar layers, which can also be linked to the transition of gas to a degenerate state. Similar assumptions existed when it was not known that novae, as well as other flare stars, form binary systems. Now it is thought that this particularity of flare stars is directly related to their explosions. Another possibility is that the second star induces the instability of the flare star.

Even more important are the flares of supernova stars. The velocity of an ejected supernova envelope reaches 6000 km/s. The mass of the ejected envelope is about one-tenth of the solar mass and the energy of the envelope can be compared to the total energy of the star before the outburst. Recent studies of supernovae in other galaxies have shown that all supernovae can be divided into two types. Supernovae of type I (among which is the outburst which gave birth to the Crab nebula) represent stars with a relatively small mass belonging to the type II stellar population. Spectral studies of ejected envelopes show that there is little hydrogen present but a relatively large amount of heavy elements, such as carbon, nitrogen, and oxygen. The masses of the ejected envelopes are rather small. Supernovae of type II exist only in plane subsystems and thus belong to the type I stellar population. Their ejected envelopes contain a great deal of hydrogen. The masses of these envelopes are much greater than those of type I supernovae. Although we have no precise indications, we can assume that the mass of an envelope ejected by a type II supernova can be greater than the mass of the Sun (up to $10M_\odot$).

Apparently type I supernovae are old stars with small mass, far advanced in their evolution, and type II supernovae are young, massive stars early in their evolution, which is proved by the great abundance of hydrogen. It is thought that type I supernovae eject an envelope, which represents only part of their mass, while type II supernovae explode entirely. The discovery of pulsars and the indentification of one of them with the supernova 1054 (in the Crab nebula) made this long-known hypothesis more credible and it is believed that the explosion process of a supernova is related to the transition of the star to the state of a neutron star after the entire consumption of the nuclear matter.

Here we end our description of non-stationary stars but we do not take leave of them. As a matter of fact, the radiation of these stars plays an important part in stellar evolution theory with which we shall now become acquainted. We shall come back to non-stationary stars of the supernova type and the very interesting T Tauri stars.

9

Protostars

All that we have studied in this book until now has led us gradually to an important problem whose understanding goes far beyond the limits of astronomy—the problem of the formation and evolution of stars. This is far from being solved and much work still needs to be accomplished. However the outline of a stellar evolution theory has appeared and a number of reliable results exist.

In this and the following chapter we shall expose the actual notions of stellar evolution—sometimes hypothetical and sometimes more promising. we can now describe a general and sufficiently probable scheme of the formation, development, and 'death' of at least ordinary stars which are not outstanding by any specific peculiarities. But how promising is this scheme?

Can we asume that the general scheme of stellar evolution, established in the last 15–20 years and explained below, is correct and will not change essentially in the future? Of course the details of this theory can indeed change considerably. Here we shall expose the theory of evolution and the birth of stars with many concrete details, but perhaps not sufficiently ascertained. The reader should keep this in mind. We emphasize again that the general scheme of the theory of evolution and birth of stars developed up to now should last a long time.

The stellar evolution theory uses many results of very different chapters of science: astrophysics, stellar astronomý, radioastronomy, intergalactic astronomy (studying galaxies), cosmology (studying the structure and development of the observed part of the universe), etc. The theory of inner stellar structure contributed mostly to the development of this problem. By building various stellar models and comparing them with observational data (in particular, that obtained by an analysis of the spectrum–luminosity relationship), astronomers were able to outline the evolutionary course wc shall describe here.

Unfortunately not all computations can be compared to observations—and this is the greatest difficulty of stellar evolution theory, as we shall soon see. It is interesting to note that 'infrared astronomy' and radio astronomy will also contribute to the theory of stellar evolution. We shall start with them to underline this role of observations.

The study of stellar clusters and associations has shown that the process of stellar formation takes place all the time, that it continues at the present time, and that stars are 'born' by groups, composed of several tens of stars, and in a much earlier period of development of galaxies—by much greater clusters. The origin of stars is rather long and of course it has not been possible, within the several tens of years during which astronomy has developed, to observe the evolution of stars and clusters. We have to compare clusters of different ages. We shall henceforth do this many times.

From what is a star generated? Evidently from the matter contained in very young stellar clusters or associations. Of course we are not sure that we see all that is there. We can make the same assumption as Ambartsumian, who, discovering an association in 1946, considered that there are invisible, dense clusters of matter, which he called D-bodies. We cannot pledge their existence on the basis of pure assumption, but we can start from what we really see in clusters and associations.

Observations show that all young clusters and associations are either imbedded in big masses of interstellar gas or dust (they are called gas–dust complexes) or they are close to them. According to the opinion of the majority of astronomers, we shall assume that stars are formed by condensation of matter of gas–dust complexes. We must again emphasize that although this is probable it nevertheless is only an assumption.

Be that as it may, the first step is done—the initial position is formulated. Now we can study the process of transformation of part of the gas–dust complex into a star. We shall do this theoretically and, whenever possible, we shall try to compare theoretical computations with observational data. The process of transition from a gas–dust cloud to a star with thermonuclear energy sources is called the protostellar stage in stellar evolution. This chapter will be devoted to this stage. In the following chapters we shall study stellar evolution including thermonuclear sources.

It is evident that to begin we must get acquainted with the characteristics of the interstellar medium. Astronomers know that the space between stars is filled with very rarefied gases and fine dust. On the average, in one cubic centimetre of interstellar space there is no more than one hydrogen atom and even less atoms of other chemical elements. In one cubic kilometre of this space there is no more than ten dust particles, each of about one micrometre.

In gas–dust complexes, and also in those where stars are generated, the concentration of gas and dust is thousands and sometimes millions of times greater than in an 'empty' interstellar space. Of course, on Earthly scales this matter is also extremely rarefied, and its density is much weaker than in the very best vacuum which can be obtained in laboratories on Earth. The characteristics of such a rarefied medium are quite varied, but we shall consider here only its temperature. It appears that for the condensation of interstellar medium into stars the most important factor is the thermal balance in the medium.

In order to determine the temperature of the interstellar medium we must estimate the heating as well as the cooling of this medium. The interstellar matter is heated by absorption of the light coming from stars and by collisions of particles of cosmic rays and X-rays with atoms, and is cooled by its own radiation. We must remember that the light flux from the star can also be represented in the form of a flux of quanta of electromagnetic waves where each quantum carries energy. In a hot star the energy of the quanta is on the average large, whereas in a cool star it is small. Even greater is the energy of X-ray quanta.

When an 'energetic' quantum falls into interstellar space, it can pull off an electron from an interstellar gas atom (ionize the atom) and 'push' it on a free trip in interstellar space, providing it with all the energy that remained after the electron has been torn off. It appears that in this case the electron acquires about the same energy as that of free electrons in the atmosphere of a hot star. The free and 'energetic' electron formed through ionization, when entering into collision with other electrons or atoms of the interstellar medium, transmits its excess energy. When its energy equals the mean energy of the interstellar gas particles it can again be captured by an ion (this is called recombination). Thus, the atom will remain an atom and the whole process can start all over again, but each time one quantum of stellar light disappears it turns into kinetic energy of the gas particles, i.e. it is heating it. If the interstellar gas were not cooled, its temperature would equal that of stars. X-ray quanta tear off atoms and inner electrons and heat them to even higher temperatures.

Until the middle of the sixties, astronomers thought that inter-stellar gas is heated by absorption of stellar radiation. Then it appeared that an important part is played by the heating of gas through absorption of X-rays and particles of cosmic rays with small energies (so-called cosmic rays). Until the launch of artificial satellites it was not known that many powerful X-ray sources exist. After their discovery it became evident that their role in the heating of the interstellar medium is important. There is no precise data on the existence of subcosmic rays, but probable estimations show that their amount is sufficient for an important heating of the interstellar medium, particularly in the regions where ultraviolet stellar radiation does not penetrate.

However, the interstellar gas is intensely cooled. We already know that in every atom or positive ion there are so-called energy levels. In a normal state all electrons of an atom are on the lowest, fundamental levels. If a certain determined energy is given to the electron of an atom, it will jump to a higher level (farther from the nucleus). Usually the electron will not stay for a long time on this level but will jump back to a lower level, releasing, in the form of a quantum of electromagnetic radiation, the amount of energy which was spent on its transmission to the higher level.

Atomic electrons can be raised to higher levels by different means. This can be done by free electrons travelling in interstellar space. To raise an atomic electron to a higher level, the free interstellar electron spends the greater part of its energy, which in the final account is converted into quantum energy of electromagnetic radiation. If these quanta are then absorbed by the same interstellar gas, the energy lost by the free electrons in the excitation of atoms will come back to them. However, if these quanta are not absorbed and leave the cloud of interstellar medium, then the energy spent by the free electron on the excitation of atoms is irretrievably lost. This is the cooling of the interstellar medium.

In order to help the reader understand what the rarefied interstellar medium is, we shall see how this mechanism of cooling would work in Earthly conditions. Here, also, free electrons can excite atoms by a transfer of energy. It must be remembered that an atomic electron sits on an upper level during a certain, although very short, time. During this short time, another free electron in a dense gas can collide with such an excited atom, which will push the atomic electron to a lower level, taking away the energy spent by the first electron to raise the atomic electron. As a result, during the whole process the energy passed from one free electron to another and the gas is not cooled at all. Only in the conditions of a very rarefied medium, where collisions of electrons with atoms are rare and where emitted quanta of electromagnetic radiation are almost not absorbed at all, the considered mechanism can and should effectively cool the gas.

There is one particularity in this mechanism of cooling: the process works by 'steps'. In fact, we know that the levels on which one can push atomic electrons are at determined distances and therefore for each excitation by electrons a determined quantum of energy must be transmitted. We assume that the interstellar gas is so cool that the mean energy of the electrons is smaller than the quantum energy needed for the jump of the atomic electron to the closest free level. It is clear that in this case the considered mechanism of cooling does not work: the free electrons are not capable of exciting the atoms and consequently cannot lose energy. The gas starts to heat. As the temperature increases, the energy of free electrons also grows. When it becomes comparable to the energy of the first excited atomic level, the mechanism of cooling stars and any further temperature increase comes to a stop or is inhibited.

Let us assume now that the gas is heated so rapidly that the cooling mechanism cannot cope with the dispersion of energy. Then the mean energy of the electrons will also increase and becomes greater than the energy of the level. The cooling mechanism acts as before but now what follows happens. Each time during the excitation of the atom the free electron loses only a determined amount of energy equal to the energy of the level, and therefore, if the level is low, the electron transmits to the atom only a small part of its energy. The cooling mechanism becomes less effective and cannot stop the increase of temperature. If the temperature increases so much that the free electrons are able to excite the higher atomic levels, the cooling mechanism is more effective here since, for the excitation and subsequently also the radiation, greater amounts of energy are necessary. The growth of temperature slows down or stops altogether. If the second level also is not able to stop the growth of temperature this can be done by the third level, the fourth, and so on.

The system of levels which cool the interstellar gas is rather complex. The energy is greatest in electron levels of atoms and ions. Therefore if the temperature of the gas is already high it will be cooled by excitation of the optical luminescence of most abundant atoms and ions in the interstellar space, such as carbon, nitrogen, oxygen. In helium the levels are high and are excited only at a very high temperature. The levels of a hydrogen atom are also rather high, but here it is important to note that at high temperatures the hydrogen atoms are ionized and in a 'bare' proton there is nothing to be excited.

At low temperatures the electron levels in atoms are not excited—the electrons do have not enough energy for it. Here the energy levels of molecules are more important. In this case the kinetic energy, as radiation, is not given up by electrons but by heavy atoms, ions, or even the molecules. If the gas temperature is not very high then, in collisions between molecules or atoms, oscillations are generated inside the molecules, i.e. the atoms in molecules oscillate with respect to each other. The energy of these oscillations is radiated in the form of infrared quanta. In the case of very low temperatures in collisions of molecules with atoms the rotation levels of molecules are excited. In other words, in collisions the molecules are untwisted which releases the kinetic energy of the motion of the molecule and the interchange of the atoms and the molecules. Then the rotation energy is radiated in the form of low frequency infrared and even radio quanta. Indeed, the greatest contribution to interstellar gas cooling is from hydrogen molecules H_2. Unfortunately it is not possible to detect the existence of these molecules in interstellar space by modern means, although there is no doubt about their existence.

One more cooling mechanism consists of the following effect. When a free electron passes close to a positively charged ion it slows down, radiating energy in the form of electromagnetic waves with a large spectrum of frequencies. This loss of energy, although small in absolute magnitude, can occur in the case of electrons with an arbitrary energy and therefore this mechanism cools the gas at high as well as low temperatures.

The gas can also be cooled in collisions of atoms and molecules with cosmic dust particles—small particles composed of graphite, ice, or other components—only if the temperature of the dust particles is smaller than the temperature of the gas.

There are therefore several mechanisms of heating and cooling of interstellar gas. Although these mechanisms are known, the problem is to know how to calculate the temperature of interstellar gas in different conditions. Unfortunately this is a rather difficult problem because it is not always possible to estimate correctly the relative part played by this factor. The temperature of interstellar gas has been calculated many times and often these calculations can be checked by observational data. We shall give the results of computations and describe the general appearance of temperature distribution in interstellar space.

What really is the temperature of insterstellar gas? Close to hot and bright stars there are many quanta of light whose energy is high (more precisely we must speak of ultraviolet quanta). They can almost entirely ionize all the surrounding hydrogen and therefore transfer to it a great deal of energy. The temperature here increases sufficiently rapidly and remains high, corresponding to the first levels of the most abundant atoms and ions of oxygen and nitrogen. The temperature is now close to 8–10 thousand degrees, which is much less than the temperature of hot stars of classes O and B.

Close to cool stars there can be many quanta of light, but the energy of each quantum is small. They are not able to ionize hydrogen, helium, oxygen, or even nitrogen. They can only ionize carbon, iron, and other atoms of low abundance. As a result there are many less free electrons, as they have obtained a smaller amount of energy from the quanta released by the star, and therefore the heating process is less effective. This leads to the conclusion that here the cooling process comes to a stop at the first 'degrees' of carbon and iron as well as at the ionization of hydrogen molecules. Close to hot stars these 'degrees' are passed over due to rapid heating. Thus, the temperature in the vicinity of cool stars should be low, and appears to be about 100–200 K (but not in the immediate neighbourhood of such stars). The heating by X-rays and particles of subcosmic rays can strongly increase this temperature, but, on the other hand, if there are many dust particles and molecules the temperature close to cool stars can be even less.

Now we shall consider what happens far from stars which are hot or cool. All ultraviolet quanta emitted by hot stars will be absorbed by interstellar hydrogen close to these stars and therefore will not reach the 'far' interstellar space. Consequently hydrogen will not be ionized there, the heating process will be slow, and the gas temperature will remain low—lower than in the vicinity of cool stars. At distances from hot stars which ultraviolet quanta can reach and where hydrogen ionization is important, the gas temperature increases strongly and can exceed 5000 K.

In general the situation is as follows. Each hot and bright star is surrounded by an extended region (its dimension can reach tens of parsecs) of ionized hydrogen, called HII regions. The temperature inside the HII region is 10 000 K in the central part close to the exciting star and decreases on the periphery of this region to 5000 K. The HII regions are separated from the rest of the interstellar space (denoted HI) by a rather sharp limit where the degree of ionization of hydrogen drops from one to zero and where, for this reason, the temperature decreases sharply. In the regions of non-ionized hydrogen the mean temperature is close to 100 K (about −170 °C) and in denser HI regions the temperature can be smaller, dropping to 50 K and even to 4–6 K. Here the main source of cooling is the excitation of the rotation levels of molecular hydrogen H_2. In the interior of dense regions of non-ionized hydrogen, particularly if there is also dust, practically no stellar radiation nor X-ray quanta and particles of subcosmic rays can penetrate. On the other hand, infrared quanta corresponding to the radiation of rotational transitions (i.e.

from the deceleration of molecular rotation) leave the dense regions of non-ionized hydrogen more or less freely and therefore cool them intensely.

We must emphasize that the temperature of interstellar gas cannot drop below 3 K, i.e. lower than $-270\,°C$, for the following reason. The entire space is filled with so-called remnant radio emissions, i.e. electromagnetic radiation which remained from the initial state of our universe when it was very dense and hot. Now this radiation has a temperature of about 3 K (more precisely 2.7 K) and as it penetrates literally everywhere (if there is not a very strong absorption at high frequencies) the temperature of interstellar gas cannot fall below this value. Indeed, all-penetrating, remnant radio emission will heat the gas to this temperature. The fact that in dense HI regions the temperature exceeds, by only a few degrees, the temperature of the remnant radio emission shows how effective is the cooling process of interstellar gas.

The reader should keep in mind this important particularity of interstellar gas: its temperature is not arbitrary, but is entirely determined by the characteristics of the interstellar medium and the exterior radiation. Now we must determine which parameters of the interstellar medium have an influence upon the temperature in the first place. One could think that first of all the temperature of a gas depends on its density. In fact if the gas density is small the collisions between particles are rare and the cooling process will act slowly. On the other hand, greatly concentrated particles often collide, are more often excited, and lose energy by radiation. It is true that if the gas density increases so much that the excitations of atoms are extinguished by collisions and radiation cannot carry away much energy, then the cooling process becomes less effective and the temperature begins to increase. However, this only happens at densities greater than the characteristic density of interstellar gas. In fact, there are also other conditions which affect the temperature of interstellar gas: the chemical composition, the relative portion of dust particles, the degree of ionization, and the presence of X-ray sources and cosmic ray particles. Nevertheless, if we consider homogeneous regions of the interstellar medium at a great distance from hot stars and other 'heating' energy sources, then the temperature of the interstellar gas drops gradually with the increase in density.

Since the density of interstellar gas is very small ($\rho \approx 10^{-25}$–10^{-23} g/cm^3) it is more convenient to use for our study not the density but the concentration of particles, i.e. the total number of atoms, ions, and molecules in a volume unit. We shall denote this quantity by the letter n. Since in the interstellar medium there is mainly hydrogen, $n \approx \rho/m_H$, where m_H is the mass of a hydrogen atom. Moreover, we can also determine the concentration of electrons in a unit of volume n_e. The ratio n_e/n is called the degree of ionization. In ionized hydrogen regions, HII, we have $n_e/n \approx 1$ and in non-ionized hydrogen regions, HI, this value is much smaller and depends on the total density. In dense HI regions $n_e/n \leqslant 10^{-3}$ and in rarefied regions it can even be $n^e/n \approx 0.1$.

Thus, the temperature of interstellar gas drops gradually as n increases.

112

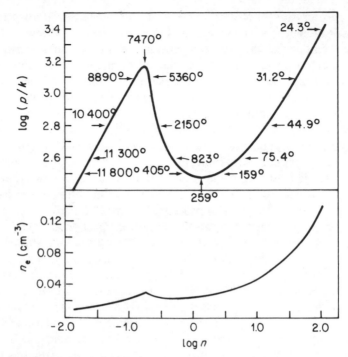

Figure 18 Graphs of the dependence of the temperature (K), pressure, and electron
concentration in the interstellar medium on the total concentration of particles

What happens to the gas pressure? This is proportional to the product of the concentration with the temperature. According to the Clapeyron law,

$$ p = \frac{A}{\mu} \rho T = \frac{A}{\mu m_H} nT = knT, $$

where $k = A/m_H$ is the Boltzman constant, $k = 1.38 \times 10^{-16}$ erg/grad. It has been also assumed here that the molecular weight of interstellar gas is close to unity.

Calculations show that at small concentrations the temperature drops as n slowly increases and therefore the pressure increases as the density increases. At greater concentrations the temperature decrease also slows down as n increases and, in this case, the pressure increases with the density increase. However, there is a certain interval with a concentration of $0.2 \leq n \leq 1\,\mathrm{cm}^{-3}$ where the temperature drops very sharply as n increases; in this case the pressure decreases as the density increases.

Figure 18 shows the dependence of temperature, pressure (upper graph), and concentration of free electrons (lower graph) on the total concentration of atoms, ions, and molecules in the interstellar matter. These graphs play an important part in the comprehension of stellar origin and we must consider this question in detail.

From the upper graph of Figure 18 we see that if the concentration of atoms is small, e.g. if $n \leqslant 10^{-2}$ cm^3 (i.e. log $n \leqslant -2$), then the gas temperature is a little higher than 10^4 degrees and the product nT, which is proportional to the pressure, is less than 3×10^2 degrees/cm^3 (i.e. log $(nT) \leqslant 2.5$). We will start from this state and will compress the interstellar cloud, increasing the outer pressure. The gas density will increase and the temperature decrease. At a concentration of 0.2 cm^{-3} the temperature will be 7500 K and the gas pressure will correspond to the product $nT = 1.5 \times 10^3$ degrees/cm^3—the curve on the upper graph of Figure 18 will reach the first maximum. What will happen if we slightly increase the pressure further? The density of the gas will indeed increase, but as the density increases in this concentration interval the temperature will drop so rapidly as n increases that the pressure, that is nT, will decrease. This means that if we compress, however weakly, the gas to a concentration exceeding $n = 0.2$ cm^{-3}, corresponding to the maximum of this curve, the gas will not only be unable to counteract the outer pressure but will continue to contract and cool along this curve up to the lower point where $n \approx 1$ cm^{-3} and $T \approx 260$ K. In other words, the state of interstellar gas in the concentration interval from 0.2 cm^{-3} to 1 cm^{-3} is unstable; it must spontaneously contract, since the effective energy emission which rapidly decreases the temperature as the density increases deprives the gas of the possibility to counteract the contraction.

Thus, a certain region of interstellar gas compressed by an outside pressure, having reached a state with $n \approx 0.2$ cm^{-1}, passes to a regime of spontaneous contraction and rapidly to a state with $n \geqslant 1$ cm^{-3} and $T \leqslant 260$ K. If the pressure is preserved, this gas will continue to contract, the temperature will decrease as before, but more slowly, and the gas will again counteract further contraction. When the concentration exceeds 10^2 cm^{-3} and the temperature drops to 15 K, the pressure will be greater than the value which existed at the point of spontaneous contraction. However, in real conditions the interstellar gas begins simultaneous contraction a little earlier as it has not reached the upper maximum of this curve.

Thus, resuming this analysis and examining again Figure 18 we can draw the following important conclusions. First, at pressures smaller than 4×10^{-20} atm (corresponding to the product $nT \approx 3 \times 10^2$ degrees/cm^3), the interstellar gas can exist only as a rarefied medium with $n \leqslant 10^{-2}$ cm^{-3} and a high temperature, $T \geqslant 10^4$ degrees. Second, at pressures greater than 2×10^{-19} atm, the interstellar gas can exist only in a dense state with $n \geqslant 10^2$ cm^{-3} and a low temperature, $T \leqslant 20$ K (if, of course, it remains non-ionized, i.e. an HI region). Third, in the intermediate pressure interval, simultaneous states (with equal pressures) of low concentrations ($10^{-2} \leqslant n \leqslant 0.2$ cm^{-3}) and high temperatures and of high concentrations ($1.0 \leqslant n \leqslant 10^2$ cm^{-3}) and low temperatures can exist. The interstellar gas in this pressure interval breaks up into two parts.

This phenomenon was visually observed a long time ago by astronomers. If we attentively observe the sky in the region of the Milky Way, where there are a great many stars, even an inexperienced eye will see the flock-like distribution: there are neighbouring regions with many stars and with few stars. This can be partly explained by the fact that in the regions showing a small amount of stars it appears that, in fact, the light of many stars is screened by absorption in different dust clouds. Thus, we can see only some close dust clouds, but a detailed study shows that all interstellar gas and all cosmic dust are distributed over the interstellar space in such a way that separate clouds (dense phase) exist in the intercloud medium (rarefied phase). For a long time astronomers did not know how to explain the fact that interstellar gas is so rarefied. At present we understand this phenomenon—it is due to spontaneous contraction of the gas which results from energy radiation. The clouds appear where, for some reason, and we shall study these reasons, the gas density has reached the 'dangerous' value close to the first maximum on the graph of Figure 18.

The second graph of Figure 18 shows that although the concentration of free electrons grows, also on average with the growth of n, this is a slow and not monotone increase. In a rarefied intercloud medium the degree of ionization is high, whereas in dense clouds the ratio n_e/n is small.

We have thus considered in detail the thermal behaviour of the interstellar gas and explained how it behaves under the action of an outside pressure. What exactly does an outside pressure mean in the interstellar medium? First, this is the general pressure of the whole gas. The interstellar gas is held in the Galaxy by the total attraction of all stars and is distributed along the plane of the entire Galaxy in the form of a cylinder of about 200 ps (6×10^{20} cm) thickness in the central parts of the Galaxy and a few times greater on the peripheries. The gas pressure should be such as to keep this cylinder from flattening further due to attraction to the symmetry plane of the Galaxy. A simple formula exists relating the gas pressure to its density and the thickness of the cylinder d which contains this gas:

$$p \approx 2\pi G\rho^2 d^2.$$

On average the density of interstellar gas and the mean density of stellar distribution are the same, about 10^{-24} g/cm^3. Substituting for this value and taking $d \approx 6 \times 10^{20}$ cm we have $p \approx 10^{-19}$ atm, which corresponds to the pressure range at which interstellar gas breaks up into two phases: a cloud and an intercloud medium.

We have not yet considered one force acting as pressure on the interstellar gas: its self gravitational attraction. In order to take this phenomenon into account we shall examine the destiny of a dense cloud with a certain mass, isolated by the instability described above from the more rarefied intercloud medium. According to formula (3) a gravitational pressure exists in this cloud tending to contract it. On contraction the gas density in the cloud increases, the temperature drops, but the gas pressure continues to grow. If the gas pressure finally appears to be greater than the gravitational one, the contraction will come to an end. However, if decreases in radiation and temperature keep the gas pressure smaller than the gravitational pressure, then contraction will go on until the density of the matter becomes sufficiently great to stop energy losses by radiation.

In order to write the numerical relations we suppose that the density and the temperature of the gas in the cloud are homogeneous, and we compare the gas pressure, calculated with Clapeyron's formula, to the gravitational pressure (3). We find that the gravitational pressure exceeds the gas pressure if the radius of the cloud satisfies the condition

$$(50) \qquad R \leqslant \frac{4\mu GM}{AT} = 6.5 \times 10^{17} \frac{\mu}{T} \frac{M}{M_\odot} \text{ cm} = \frac{0.2}{T} \frac{M}{M_\odot} \text{ ps.}$$

If on contraction of the cloud the temperature changes so that the inequality (50) is preserved, the gravitational forces will continue to contract the cloud. The condition (50) is, of course, rather strict. We already know that the minimum temperature of interstellar gas is not lower than 3 K. On the other hand, the density of interstellar gas with a low temperature cannot be very large, otherwise the cloud would become opaque and radiation would stop cooling. Observations show that in the coolest gas–dust clouds the concentration of particles is 10^3–10^4 cm^{-3}, i.e. the gas density is 10^{-21}–10^{-20} g/cm^3. Setting the mass of the cloud M we find the radius of the cloud with the following density: $R \leqslant (0.1 \text{ to } 0.2) \times (M/M_\odot)^{1/3}$ps. Using this value in (50) and assuming $T \approx 3$ K we find that a cloud with a mass of the order of the solar mass, having a maximum optical density and a minimum possible temperature, will contract under the action of gravitation.

The inequality (50) can be satisfied more easily if we consider clouds with greater masses. Let us increase the mass of the cloud 10^3 times. Then its radius will increase ten times and the 'reserve' of inequality (50) will now be a hundredfold greater. In other words, inequality (50) will be realized for such a cloud even if the gas temperature inside it is 300 K.

The following pattern of phenomena appears. The mean pressure in the interstellar gas is such that it breaks up into two parts. Dense clouds with low temperatures are generated. If the mass of these clouds is small, comparable with the mass of single stars, then the gravitational forces in them are either small compared to the gas pressure or they are more or less balanced by the pressure.

However, if clouds are generated with great masses, a few thousand times the solar mass, then self gravitational forces appear which contract the cloud. The increase in density is at first accompanied by a temperature decrease and therefore the contraction is accelerated. However, the entire cloud is not totally contracted as a single body. With a density increase and a temperature decrease the criterion (50) is always fulfilled for smaller masses. This means that if inside the cloud (e.g. close to its centre) a small denser part is formed, it will start contracting independently and more rapidly than the rest of the cloud. The cloud will finally break up into parts, and for each one of them the condition (50) will be realized. The disintegration into such fragments will come to a stop at the moment when the gas density is 10^{-21}–10^{-20} g/cm^3 and its temperature 4–6 K. In other words, the initial cloud with a mass of 10^3 solar masses disintegrates into several hundreds of fragments with masses of the order of the solar mass—the germs of future stars.

We can now understand why stars must generate in groups: otherwise it is difficult to satisfy condition (50) in real conditions of the interstellar medium. Condition (50) plays a very important role in astronomy. This had already been found at the end of the last century by the English astronomer D. Jeans, and has since been called the 'Jeans instability criterion'.

To make the formation of big clouds in the interstellar medium possible there should be some kind of outside pressure—something like an initial contraction. As a matter of fact, observations show that new stars do not appear everywhere in interstellar space, but only in determined places with conditions favouring the appearance of massive gas clouds. In particular, initial contraction takes place in the spiral arms of the Galaxy. Unfortunately we do not have the means to explain the nature of this phenomenon. Spiral arms are great waves in the plane of the Galaxy in which the contraction of the gas happens. In favourable conditions the gas flowing into a spiral arm is contracted by more than ten to fifteen times. This is enough for the generation of great dense clouds with a mass of about a thousand solar masses.

There are also other causes favouring the gas contraction. We know that in regions of ionized hydrogen the temperature is a hundred or more times greater than in regions of non-ionized hydrogen. Consequently, how many times greater is the pressure if their densities are the same? Let us imagine a cool gas–dust cloud near an HII region—the pressure of this region will lead to a complementary contraction of the cool cloud. For this reason the formation of HII regions in gas–dust complexes helps to create new stars. HII regions appear with outbreaks in these complexes of very young hot stars. Here something acts as the mechanism of a chain reaction: the generation of first stars implies the generation of other stars.

Let us consider that after all fragmentation of the interstellar medium, first into big gas clouds and then into single fragments of small mass, densities and temperatures are reached when the inequality (50) is satisfied for clouds with masses close to the solar mass. This means that germs of single stars are formed. Unfortunately it is difficult to determine these conditions with enough precision. We have already seen that densities of about 10^{-21}–10^{-20} g/cm^3 and temperatures of 4–6 K are sufficient for further contraction of protostars and certainly these observational data result from measurements. It would be best to observe the radiation of such star embryos but to do this we must pass to the infrared region of the spectrum. We have already noticed that the low gas temperature in such clouds is maintained by radiation of hydrogen molecules. The most intense radiation is emitted under the form of infrared quanta with a wavelength of 28 μm. When we are able to observe the sky with sensitive infrared telescopes we will obtain reliable data on the star embryos!

In recent times one more possibility to observe these objects has appeared. The development of radio astronomy led to the discovery of numerous simple and complex molecules in interstellar space. The majority of them behave in an ordinary way, absorbing and emitting radiowaves as in laboratories on Earth. But water H_2O and hydroxyl OH molecules show a quite uncommon behaviour; in certain cases they show such intense radio emission that could appear only in masers. The principles of a maser are probably known to the reader. In laboratories on Earth maser radiation is emitted by special generators manufactured under strict observance of precision criteria. Apparently nature can also emit masers in natural conditions. Maser sources with radio emission of H_2O and HO molecules are effectively found in the cosmos and observations show that they are situated in places where new stars appear. It is possible that the birth of a star announces itself by the inclusion of a corresponding maser. We are not able to interpret properly the optical data of maser source observations and we still do not have reliable data of infrared studies. Therefore we shall continue on the sole basis of theoretical considerations.

Let the embryo of the star be in a critical state in which inequality (50) is fufilled at the upper limit. Since the gravitational pressure is always greater here, the contraction will continue. In Figure 18 we saw that with a density increase the temperature drops. As we are now beyond the limits of application of this graph it is not very clear how the temperature will evolve on further contraction. To simplify this question we shall consider that, at least in the initial stage of this contraction, the gas temperature in this cloud remains constant. On contraction, the gravitational pressure will increase more rapidly than the gas pressure. The contraction becomes more and more accelerated. We can now consider that almost nothing impedes the contraction and the star embryo, which we shall now call a protostar, will in a way decay, i.e. contract with the free-fall velocity of a body on its surface.

We will try to estimate the time for the contraction of a protostar, with the following considerations. On the spherical surface of the protostar the acceleration due to gravity is $g = fM/R^2$. Under the action of this acceleration the surface falls towards the centre. If the motion is all the time at a constant acceleration, then according to the well-known formula of mechanics the distance after a time t would be determined by the formula $S = \frac{1}{2} gt^2$. In fact, the surface of the protostar always moves with a greater acceleration, but we shall not consider this. Later we shall determine the time the surface of the protostar needs to fall a distance equal to the radius of the protostar. In fact, the protostar does not collapse until reduced to a point, but with the above-mentioned assumption we partly compensate for the error linked to the hypothesis of constant acceleration.

Thus, assuming $R = \frac{1}{2} gt^2$ and introducing here the expression for g, we can find the collapse time for a protostar:

$$(51) \qquad t = \sqrt{\left(\frac{2R^3}{fM}\right)} = \frac{1}{\sqrt{\{(2\pi/3)f\bar{\rho}\}}}$$

It is interesting to note that the collapse time depends only on the initial mean density of the protostar. We substitute formula (50) into (51), determining the radius of the protostar at which the collapse starts. We then have

$$(52) \qquad t = \left(\frac{5\mu}{AT} \right)^{3/2} fM = 6 \times 10^7 \frac{\mu^{3/2}}{T^{3/2}} \frac{M}{M_\odot} \qquad \text{years.}$$

In particular, at $T = 50\,\text{K}$ a protostar with the mass of the Sun will contract within ten thousand years. A contracting protostar emits energy in the infrared range; we shall try to estimate its luminescence. Although we still cannot observe this radiation it is interesting to know on what we must rely. In Chapter 4 we have already considered the case of radiation of a star which is contracting and we obtained formula (11) defining the contraction time for a given luminosity. In a star luminosity is determined by the filtering of energy and therefore it is given in fact. In a protostar the collapse time is given, i.e. the time of energy emission (formula 52) which now determines the luminosity; therefore in order to find the luminosity of a protostar at the stage of its collapse we must transform formula (11), expressing L by M, R and t. We obtain

$$(53) \qquad L = \frac{fM^2}{2Rt}$$

We know the mass M of the protostar. The time of collapse t can be determined from formula (52). What happens to R? In the process of collapse of a protostar R decreases rapidly. It is evident that as long as R is big the luminosity of the protostar is small. Consequently the most favourable case for observations is during the final stage of the collapse of a protostar when R diminishes to the smallest possible value and L increases correspondingly. Let us try to determine this radius.

We shall once again consider the collapse of a protostar. The emitted gravitational energy will heat the gas or dust and this heat will rapidly be radiated. The gas and dust can again 'receive heat'. If too much gravitational energy is freed, strongly heating the gas and dust, the heat can no longer manage to radiate and brings about dispersion of the dust, dissociation of the molecules, and ionization of the atoms. In the place of gas and dust a plasma appears in the protostar which at low temperature radiates weakly. Thus, the collapse of a protostar comes to an end as soon as the gravitational energy reaches the thermal energy necesary to the transformation of the whole mass of the protostar into a plasma. We denote the energy needed to transform one gram of initial matter into plasma by the letter I. Thus, for the transformation of the entire protostar into plasma IM ergs are needed. Since the gravitational energy of a protostar is fM^2/R, we find by comparison of both values that

$$(54) \qquad R \approx \frac{fM}{I} \approx 80 R_\odot \frac{M}{M_\odot} .$$

To obtain the numerical value of R we took $I = 2.5 \times 10^{13}$ erg/g. This rough estimate has been obtained in the following way. In the interstellar medium hydrogen is the most abundant element. In the cool HI region H_2 molecules are the most abundant (3×10^{23} for one gram). For the dissociation of one molecule 7×10^{-12} erg of energy are needed. Each hydrogen atom must be ionized and for this we need 22×10^{-12} erg. Altogether, for one molecule we need $\sim 5 \times 10^{-11}$ erg and for one gram $\sim 1.5 \times 10^{13}$ erg. If we consider that interstellar space is composed not only of pure hydrogen but also of other atoms, in particular helium which is more difficult to ionize, it is better to overestimate I, taking 2.5×10^{13} erg/g.

Thus, a protostar with a mass of the Sun collapses to a radius approximately eighty times greater than the radius of the Sun. In more massive protostars the minimum radius of collapse is also bigger. Now we substitute (54) and (52) into (53) and obtain the luminosity

$$(55) \qquad L = \frac{I}{2f} \left(\frac{AT}{5\mu} \right)^{3/2} \approx 0.002 L_\odot \left(\frac{T}{\mu} \right)^{3/2} \qquad \text{erg/s}$$

This is a curious result—the luminosity of a protostar at the final stage of free fall does not depend on its mass and is determined only by the temperature of the gas. What will the value of T be here? It is no longer the initial temperature of the interstellar gas since in the process of collapse the temperature in the interior of the protostar continues to grow. Apparently the temperature should be chosen to be close to the temperature at which the dissociation and ionization of hydrogen starts. We take $T \approx 10^4$ degrees. Assuming also $\mu \approx 1$ from (55) we obtain $L \approx 2 \times 10^3 \, L_\odot$. Consequently a bright flare occurs for the protostar, although only for a short time—possible for a few years or even less.

Immediately after the flare the luminosity starts to decrease and the protostar becomes opaque. However, the contraction does not stop as now the energy released in the star serves to heat its interior. A large temperature decrease occurs and therefore some energy transfer mechanisms could 'be included' from the inside of the protostar to the exterior.

Astronomers assumed that during the contraction of a protostar the energy generated is transmitted through its thickness in the same way as in an ordinary star. Let us also assume that this is in fact so. We can then use formulae (8) and (9) to determine the opacity of protostars. From this several conclusions immediately follow. First, the luminosity of such a protostar is determined by its mass. Second, if we trace the position of protostars on the spectrum–luminosity diagram, then on contraction the point representing them would shift along a horizontal line (at constant luminosity) from right to left—as it contracts the radius decreases and the surface temperature increases. Third, as the temperature increases in the interior of protostars the opacity decreases and this should increase, a little, the luminosity—the evolution sequence increases a little. These were indeed the notions of protostellar evolution described in the first edition of this book (see Fig. 19 of the first edition.

However, as has been shown by Hayashi, in a contracting opaque protostar energy is not transmitted by radiation but by convection. Calculations confirm this effect but it can also be easily explained. First, convection can transmit a greater amount of energy than radiation and therefore favours a more rapid contraction of the protostar. Second, the dissociation of molecules and ionization of atoms in protostars decreases their adiabatic curve index. We studied this phenomenon in detail in the preceding chapter while considering the causes of oscillation in pulsating stars. The same thing also occurs in protostars, only there appears to be instead of small amplitude oscillations violent convective motions which envelope the entire protostar—we can say that it is 'boiling'.

Thus, the energy of contraction is transmitted from the interior of opaque stars to their surfaces by convection. However here their energy must be converted into radiation otherwise it cannot leave the protostar. In this way radiation still plays a part in the contraction of a protostar—it determines the conditions on its surface. But how? This can be easily explained by analogy with the study of stars where energy is transported by convection to the outer layers, where it must also be converted into radiative energy before leaving the star. We know that convection in surface layers occurs only in cool stars. This is a characteristic feature of red giants where an important part of the stellar volume is convective, continuing up to the very surface. In all such stars the surface temperature is close to 3500 K. This means that at such temperatures, or slightly lower, the transformation of convective energy into radiative energy takes place and that precisely this temperature should be at the surface of opaque protostars.

We must note, however, that the surface temperature of convective stars depends only slightly on the mass and the radius. This dependency has a complex nature: T_e can either increase or decrease as M and R increase. For precise calculations this fact must be taken into account, but for the explanation of the protostellar evolutionary pattern it is sufficient to neglect the change in T_e.

We can imagine the following pattern. As soon as the protostar becomes opaque its inner temperature begins to increase—first slowly and then more rapidly. Convection appears in the protostar to rapidly take up all its volume. The temperature continues to increase in its centre but the temperature on the surface reaches three thousand degrees and subsequently hardly changes. When the temperature at the surface of the protostar comes close to this value, the luminosity of the star will, after a certain drop, again reach a maximum—the convection carries the greatest amount of energy from the contracting protostar. Later the luminosity will again start to decrease since, according to (1), with the decrease in the radius and a constant surface temperature the luminosity can only decrease. It is rather difficult to determine the luminosity of a protostar at the moment of its maximum flare, i.e. at the moment when the convection takes up the whole star. For this we need detailed calculations but unfortunately the available information is not very reliable. However, we can also obtain a simple estimate if we consider that during the convective stage the protostellar radius would not change very much—in other words, we can take the value (54) for the protostellar radius at this moment. Then, using formula (1) and considering that the effective temperature of the protostar is about two times smaller than the effective temperature of the Sun we obtain

$$(56) \qquad \frac{L}{L_\odot} + \left(\frac{Te}{T_\odot}\right)^4 \left(\frac{R}{R_\odot}\right)^2 \approx 400 \left(\frac{M}{M_\odot}\right)^2 .$$

This is really a bright flare, although also of short duration (a few years).

Later the luminosity of the protostar decreases in proportion to the decrease of the square of the radius. The duration of contraction can be determined as before using formula (11). We have

$$(57) \qquad t = \frac{fM^2}{2RL} = \frac{fM^2}{8\pi\sigma T_e^4 R^3} = \frac{\bar{f}M\rho}{6\sigma T_e^4} \approx 8 \times 10^7 \bar{\rho} \; \frac{M}{M_\odot} \qquad \text{years.}$$

It is interesting to note that here the duration of contraction is proportional to the mean density whereas in a transparent protostar the duration of contraction is proportional to $1/\sqrt{\bar{\rho}}$.

During this contraction the central temperature increases quickly. It finally reaches a value at which thermonuclear reactions start. The protostar continues to contract, but when the thermonuclear reactions are in a state to produce enough energy to maintain a high temperature the protostar 'sits' on the main sequence and turns into an ordinary star. Its destiny will be studied in the following chapter.

Thus during contraction of a protostar two bright flares occur, the first at the end of the 'transparent' stage, when the luminosity reaches several thousands of solar luminosities independently of the protostellar mass and when the greatest part of radiation takes place in the far infrared region of the spectrum, and the second at the moment of establishment of convective energy transfer, when the luminosity is smaller, about $400(M/M_\odot)^2 L_\odot$, and the radiation corresponds to the surface temperature of a red star. The time interval between these flares is small (a few years) and it is possible that in fact they merge. Be that as it may, in the protostellar stage there exists a certain interval of very strong protostellar luminosity. Observations of these flares would be an important confirmation of the theory.

The formulae given here indicate correctly all characteristic features of protostellar evolution, but of course they are approximate. More precise calculations can be done and have been done many times. We must note, however, that the precision of these calculations is in a certain measure illusory since many parameters are not known with any great certitude.

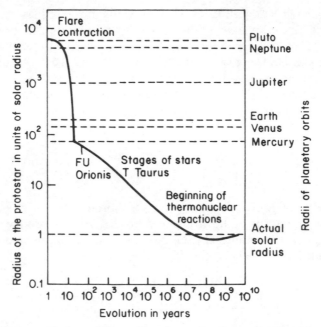

Figure 19 Variation with time of the radius of a contrasting protostar. A luminous flare corresponds to the stage of rapid contraction (left-hand side of the graph)

As an example we shall give some results of such calculations. Figure 19 shows the variation of the radius of a protostar of a mass equal to the solar mass with time. To illustrate this example we also give the orbital radii of planets in the solar system. The region of rapid contraction on this graph corresponds to the period immediately before the flare and lasts for all of about ten years. The more gently inclined part of the curve shows the contraction during the convective stage of protostellar evolution.

Figure 20 represents the displacement of a protostar in a spectrum–luminosity diagram. At the beginning there is rapid contraction in the transparent stage (the dashed line at the extreme right-hand side of the graph). Here the uncertainty is linked to the value of the gas temperature and therefore two arrows are marked. Then the protostar becomes opaque, shown by a rapid drop of the dashed line on the diagram. Further, a large shaded area represents a sharp 'expansion' of the star at the instant of a flare ('boiling') of the protostar. Finally, the decrease in luminosity in the convective stage of contraction is shown up to the transition to the main sequence.

We have already noted that all these effects occur mainly in the infrared region of the spectrum. In recent times it has been possible to make the first infrared stellar observations. It appears that there are a few very bright objects with surface temperatures of about a thousand degrees. One of these objects was identified as the well-known variable star R Monoceros while another object appears to be a punctual source in the Orion Nebula. Perhaps these objects are protostars at the instant of a flare. Figure 20 shows such a hypothesis.

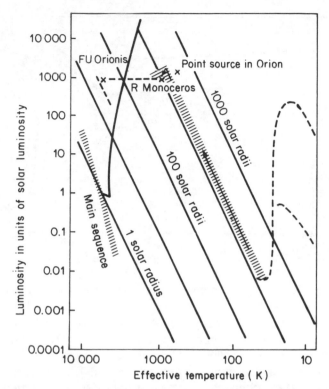

Figure 20 Spectrum–luminosity diagram with the evolution track of a protostar. A large radius corresponds to the transition moment in the convective stage. This has been done for a better agreement with observations of infrared objects (graph by Hayashi)

The case of the variable star FU Orionis is also interesting. In 1936 this star suddenly flared and since then continues to radiate with approximately the same intensity. Its radius is about twenty to twenty-five times greater than the solar radius. This is also a flare but with a transition to a higher surface temperature and a subsequent slower contraction. This hypothesis is also represented on Figures 19 and 20.

If observations of protostars in their early evolutionary stages are a matter for the future, protostars in the stage of convective contraction should be observed now.

How can we distinguish such protostars from stars with the same surface temperature? They should have more intense convection, but of course we do not see this. We can expect that the violent 'boiling' of protostars should bring about a greater activity of their atmospheres; we can assume that on the surface of protostars flares and outbursts of different magnitude should be observed. Contracting protostars resemble non-stationary flare stars. If we consider that they could not move far from their original place in the gas–dust complexes and that such objects should be seen in very young clusters and associations, then we can immediately select protostars.

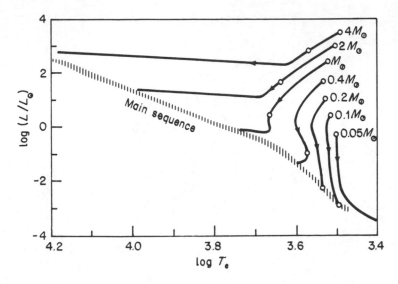

Figure 21 Evolution tracks of protostars contracting towards the main sequence

As a matter of fact in the early forties A. Joy discovered stars that were later called Taurus stars. V. A. Ambartsumian showed that these stars are always grouped in associations (called T-associations). Close to T-Taurus stars is always an interstellar medium. The features of T-Taurus stars are very characteristic. They often change their brightness several times and the increase or decrease in brightness can last for hours. However, there are also frequent periods without any change in the brightness of the star. In these stars sometimes lines appear due to the ejection of gas clouds from the surface layers of the star at velocities reaching 200–300 km/s. An important fact is also that these stars are rich in lithium: it is about fifty to four hundred times greater in abundance than in the Sun. This, together with the absence of high temperatures in their interiors necessary for thermonuclear reactions, is evident proof of the youth of these objects. With all these data it is nowadays admitted that these T-Taurus type stars can be considered as protostars in the convective contraction stage.

Until now we have considered the evolution of protostars with a mass close to the solar mass or with smaller masses. In more massive protostars, before their transition to the main sequence and the start of thermonuclear reactions, the convective transfer of contraction energy is converted into radiative transfer. This is due to the fact that in massive stars the gas is entirely ionized, which on the one hand increases the adiabatic curve index (diminishes the polytrope index) and on the other hand decreases the opacity. For this reason when massive protostars contract their 'vertical drop' on the spectrum–luminosity diagram with convective transfer is replaced by a horizontal shift (which we mentioned at the beginning of this chapter) with a radiative energy transfer. This is schematically represented on Figure 21. With formula (9) we can obtain the luminosity of a protostar at this stage of contraction; it is not much different from the luminosity of a star with the same mass, as can be seen in Figure 21.

On this graph, constructed by the Japanese astrophysicist Hayashi, the circles denote the moment of generation of a radiative nucleus in massive stars or the moment of transition of the main sequence in stars with a mass $M \leqslant 0.25M_\odot$ where, as we know, the radiative nucleus is not in general formed. In protostars with $M \leqslant 0.05M_\odot$, during the contraction process the central temperature remains so low that nuclear reactions are not included, the contraction stops with the degeneration of the electron gas in the centre of the protostar, and such a star does not arrive at the main sequence.

On contraction of a protostar with a mass equal to the solar mass a central radiative nucleus is generated before the transition of the star to the main sequence. Precise calculations made for the Sun show that inside the 'protosun' a radiative nucleus is generated when the protosun contracts to a dimension two times greater than the actual solar radius. The luminosity of the protosun was then equal to $1.5L_\odot$. As the contraction continues the luminosity decreases to a value of $0.512L_\odot$; in this case, in the greatest part of the protosun energy is already transported by radiation. The variation of the absorption coefficient during the heating process of the protosun on contraction brings about, according to (9), a subsequent luminosity increase reaching a magnitude close to $L_\odot$.

Using formula (11) we can easily estimate the duration of contraction on the horizontal part of the evolution path:

$$(58) \qquad t = 2 \times 10^4 \frac{\varkappa}{\mu^4} \left(\frac{M_\odot}{M} \right) \left(\frac{R_\odot}{R} \right) \qquad \text{years.}$$

The characteristic contraction time depends on the radius of the protostar and rapidly increases with the decrease in the radius during the contraction process. This means that such protostars rapidly cross the right-hand side of the diagram in Figure 21 but slow down as they approach the main sequence.

Formula (58) is the result of an approximate estimate. The contraction time of protostars can be calculated more precisely, as shown by Figure 22. In order to explain the meaning of this graph we shall study the following case. We assume that in a given gas–dust cloud contracting stars of different masses were simultaneously formed and we trace on a Hertzsprung–Russel diagram the position of these stars during equal time intervals. Then lines will appear on this diagram which decribe sequences of protostars of the same age. These lines are called isochronisms. Figure 22 shows isochronisms of contracting protostars. The first line on the upper right-hand side corresponds to the position of protostars 10^4 years after the beginning of the contraction, the second line (if we continue towards the lower left-hand part) after 10^5 years, and the third line determines the position of protostars after 10^6 years. The following two lines represent protostars after 10^7 and 10^8 years respectively.

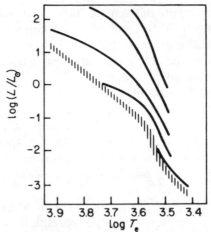

Figure 22 The isochronisms of protostars contracting towards the main sequence. The curves represent stellar ages (from above to below) of 10^4, 10^5, 10^6, 10^7, and 10^8 years

This graph also shows that massive protostars rapidly skip the region above the main sequence and stay there during several millions of years after their generation. Protostars with small masses pass to the main sequence after hundreds of millions of years.

From Figures 21 and 22 we can see that massive protostars in this stage of evolution belong to the same region as red giants. It is therefore possible that some parts of the red giants, observed in young stellar clusters, are in fact still protostars. However, their number should be small since protostars rapidly 'skip' this region because of the short duration (58).

In fact, an analysis of spectrum–luminosity diagrams of young clusters allowed us to discover clearly defined sequences of red giants, as well as a few stars dispersed over the diagram in this region—apparently protostars of large mass. It is also possible to observe quite young stellar clusters which probably have not yet managed to settle on the main sequence; their spectrum–luminosity diagrams are the same as the isochronisms of Figure 22. An example of such a protostellar cluster is the object known by astronomers as NGC 2264. Its spectrum–luminosity diagram passes near the isochronisms corresponding to the age of 10^6 years.

We must note that very bright stars, particularly stars of classes O and B, occur, as a rule, in clusters as well as in associations (O-associations); moreover, O- and T-associations are often adjacent. From condensations of gas–dust complexes stars are apparently formed with different masses. We must finally note that in big and dense gas–dust complexes nebulous objects are often observed, sometimes appearing as dark, round spots on the shining background of bright nebulae (globules) and sometimes as bright spots on a dark background of absorbing nebulae.

Of course optical data actually allow us to understand better stellar formation. We know that the process of stellar formation is everlasting and presently in the Galaxy many new stars are being generated (probably on average about ten stars in one year). What actually takes place is only the remains of stellar formation. Presently in the Galaxy the interstellar gas represents only 1–2 per cent. of its total mass; the rest has already managed to condense into stars. The process of stellar formation at early evolutionary stages of galaxies and of the Universe in general was much more intense. Unfortunately, this early process of stellar formation is even more difficult to observe; apparently to do this one must study other galaxies where for some reason the stellar formation process has slowed down.

However, many conclusions can be drawn from observations of our Galaxy, namely from the study of the composition and distribution of old stars in it. We recall (see Chapter 1) that in our Galaxy two basic types of stellar populations exist: type I to which the sun belongs and type II to which stars of spherical systems and globular systems belong. Where is the difference? As we can see from Table 2, one of the fundamental differences lies in the chemical composition. The heavy elements, of large magnitude Z, decrease with the transition from type I to type II populations. This fact already verifies the difference at the moment of their formation. The second essential difference lies in the distribution of the stars in space. This can be better judged not by their instantaneous position but by their motion in the Galaxy. Stars of type I population move on almost circular orbits near the galactic plane while stars of type II population move on strongly elongated orbits. These orbits are situated so that the corresponding stars pass close to the centre of the Galaxy and then move far from it at great distances, rising at the same time high above the galactic plane.

Stars of extreme type I population move in the same way as clouds of interstellar gas move at the present time. We can therefore draw the following conclusion: stars of type I population were formed of interstellar gas relatively recently, since from the moment of their formation until now the composition of the interstellar gas has hardly changed. An argument in favour of this assumption is that the chemical composition of the interstellar gas and of extreme type I population stars is almost the same.

In terms of age we can think that stars of extreme type II population were generated first, then stars of type II population, stars of old type I population, stars of type I population, and finally the extreme type I population (see Table 2). There are many good reasons for drawing this conclusion.

First, the chemical composition should change with the age so that the younger the star the greater the value of Z. In fact we know that heavy elements are formed in stars during thermonuclear reactions. Certain stars lose mass which mixes with the interstellar medium. This means that as time passes the interstellar medium should be enriched by heavy elements from the matter which managed to condensate into stars and then was again ejected into interstellar space. While the star remains on the main sequence only the helium abundance changes; the magnitude of Z does not vary. Heavy elements are formed only in the last stages of evolution (see the following chapter). Therefore the value of Z which is observed in stars on the main sequence corresponds to the value at the moment when the star is generated.

Second, the various movements of stars of different ages in space can also be easily explained. We shall consider that at the beginning of evolution of the Galaxy, when the greatest part of the matter was still a gas, it was distributed in the form of a more or less homogeneous sphere or a somewhat flat ellipsoid. As we know, gas disintegrates into clouds. The clouds moving in space collide, are heated, and radiate the excess energy. If the gas exists for a sufficiently long time it must lose a great deal of kinetic energy of the clouds by radiation. However, the rotation does not disappear in this case. This means that initially the spherical rotating system of interstellar gas gradually decreases its thickness but the diameter does not decrease, otherwise the rotation would not be maintained. The system of rotating interstellar gas becomes a more and more flattened spheroid and finally turns into a flat disk. During this time stars are formed from the interstellar gas which maintains the motion the interstellar gas cloud had at this time. In fact, contrary to big interstellar gas clouds, stars of small dimensions do not collide and do not lose their kinetic energy.

This explains an important optical observation: the younger the star, the closer to the galactic plane it moves. Consequently all particularities of division of stellar subsystems into spherical, intermediate, and plane ones, taking into account the changes in chemical composition, can be easily explained, at least qualitatively.

Now we shall come back to the problem of the mechanism of condensation of protostars from the interstellar medium. We have already studied the formation of the youngest stars and considered the physical state of interstellar gas at the present time. This gas is rich in heavy elements and, what is very important, has a rather abundant addition of cosmic particles. Its essential role in the whole problem is that H_2 molecules are formed on the dust which very effectively cools the interstellar gas by shading the innermost dense clouds from exterior radiation.

At the initial state of the interstellar gas there were little or no heavy elements. Therefore there was no cosmic dust and consequently the formation of H_2 molecules was difficult. Of course interstellar H_2 molecules can also be generated without the help of cosmic dust, but in a much smaller quantity. All this shows that at early evolutionary stages of the interstellar medium the cooling processes are much weaker. If in the actual interstellar gas the known cooling processes can decrease its temperature to 4–6 K, then in the 'earlier' interstellar gas the temperature scarcely drops lower than 200–300 K.

This conclusion slightly changes the pattern of the stellar formation process at the first stages. The general scheme remains essentially the same and all the formulae given above remain valid. Only now greater values of temperature must be used in our estimations. This means that during the first stage of evolution of the Galaxy the formation of massive stars and big clusters is more probable. Massive stars formed in the initial stage evolve rapidly and explode, 'polluting' the interstellar medium with heavy elements. Presently no initial massive stars are left and those which remain are a little later, less massive stars, slowly evolving. The big clusters of old stars are the so-called globular clusters also composed of stars with a small content of heavy elements. This is the approximate scheme of the gradual formation of protostars and stars from the interstellar medium at different stages of its evolution. It is, however, only a qualitative pattern. Some computations have also been done.

In this model many things are not yet clear but on the whole the outlines of the scheme of protostellar and stellar formation are sufficiently precise. We hope that in the future this scheme will be conserved, although in the process of further study many new concepts will be clarified.

10

Stellar evolution

We have studied the entrance into the main sequence of a protostar and its transformation into a star. Now we shall consider the evolution of stars. This is at the same time both an easier and a more difficult task. It is easier because stars can be observed and with the optical data we can reject bad theoretical models and can use observations to find different rules determining stellar evolution. It is more difficult because, in contrast to the case of protostars, we need precise data on the temperature and density of the stellar matter, on its chemical composition, on the distribution of mass inside the star, on its rotation, etc. In fact, the sensitivity of thermonuclear reactions to the temperature does not permit us to use simple numerical estimations such as those we used in the preceding chapter. The chemical composition determines the opacity and the role of different thermonuclear reactions and its influence on the evolution of a star is very strong. Finally, the most important point is probably that the evolution of a protostar is almost independent of its inner structure; in the case of stars the situation is very different, as we shall soon see. A conclusion can be drawn: for the study of stellar evolution we must calculate a large number of precise stellar models. Without the help of computers this cannot be done.

However, first a qualitative image—the essential nature of stellar evolution—can be obtained with the help of elementary notions based on stellar physics data which the reader has seen in the preceding chapters. We shall therefore first describe this qualitative picture and then give the concrete data obtained from results of computer calculations.

On contraction the central temperature of a protostar increases according to formula (5) (or formula (7) if we take $\mu = 0.6$). The thermonuclear reactions which consume hydrogen start at a temperature of eight million degrees. Introducing this value in (7) we obtain that at the moment of 'inclusion' of thermonuclear reactions the protostellar radius was $R = 1.7R_\odot M/M_\odot$. This value is slightly larger than its radius on the main sequence but the 'inclusion' of thermonuclear reactions is not able to stop the contraction. Only after equilibrium between the energy generation and its evacuation according to the mass–luminosity relation does the star finally 'sit' on the main sequence. In stars of small mass the 'inclusion radius' of thermonuclear sources is also small—if it is smaller than the radius of a white dwarf with the corresponding mass then hydrogen will not burn there.

If we mark on the spectrum–luminosity diagram the points describing the situation (i.e. luminosity and radius) of stars of different mass where the thermonuclear reaction begins to set in, we will obtain a line called the 'zero age' main sequence. Its position depends on the chosen chemical composition. For type I population stars its position coincides with the lower edge of the observed main sequence. In type II population stars with a small amount of heavy elements the zero age sequence coincides with the subdwarf sequence. The mass–luminosity diagram (Figure 11) also shows the zero age sequence (lower curve).

In a star on the zero age sequence hydrogen starts burning—in the proton reactions for low mass stars and in the carbon–nitrogen cycle for massive stars. What happens then to these stars? We shall try to imagine this using formula (9). As the hydrogen burns, the molecular weight increases, and according to (9) this should lead to an increase in the luminosity. However, in order to oppose the increased luminosity with a lower hydrogen content, the star must increase its temperature. This in turn leads to a decrease in the opacity and consequently to a further increase in the luminosity (always according to formula 9). Thus, when hydrogen burns the luminosity of the star increases, a condition represented on Figure 11 by vertical arrows.

The variation of the stellar radius can be estimated using formula (5). The temperature T_c only slightly increases with time, the thermonuclear reactions being very sensitive to the temperature. A very small increase in the latter is enough to increase the energy emission and this leads, according to (5), to a small decrease in the stellar radius. On the other hand, as the hydrogen consumption continues the molecular weight increases in the star and this condition, according to (5), brings about a more important increase in the stellar radius. Precise computations show that, although the luminosity increases, the increase in the radius is more rapid, so that the surface temperature drops. Of course these precise calculations depend on the assumption of stellar matter mixing. If we assume, as it is done nowadays, that the mixing is not very strong and that hydrogen burns mainly in the centre, then the surface temperature drops. However, if we allow for strong mixing in the star (such that the hydrogen content decreases over the whole star), then we can also obtain an increase in the surface temperature. Because of the optical factor the choice between these two assumptions goes in favour of the first one—we shall come back to this problem later.

Since the molecular weight changes only a little even after the entire consumption of hydrogen, (see Chapter 2), we can conclude that within this time the luminosity and the radius of the star also change very little, i.e. the star will remain within the limits of a relatively narrow band of the main sequence all this time. To be precise, the small variation in the luminosity and surface temperature during the whole time of hydrogen burning defines the existence of the main sequence as a well-expressed rule.

Let us continue. Since the hydrogen content in the star is proportional to its mass and the velocity of its radiation (i.e. luminosity) is proportional to the mass cubed (or even at a higher degree), it is clear that the hydrogen will burn much more rapidly in massive and hot stars than in small and cool ones. Figure 11 shows the time of hydrogen burning in stars of different masses. In a star with a mass fifteen times greater than the solar mass the hydrogen burns within ten million years and in a star with a mass one-fourth of the solar mass, within seventy milliard years. It is known that the part of the Universe which we observe exists for about ten milliard years. From this it follows that even the oldest stars with masses smaller than the solar mass would not have time to 'burn up' their hydrogen and leave the main sequence. Moreover, the smallest stars would not even have left the zero age sequence. The dashed line on Figure 11 indicates the positions of stars with small masses during 10^{10} years after the beginning of hydrogen burning and confirms what has been said above.

Thus the evolution of stars with masses inferior to the solar mass, i.e. the majority of stars, is simple and uninteresting. Once these stars have arrived on the main sequence (type I population) or on the subdwarf sequence (type II population) they remain almost in the same place. The brightness increases a little, the surface temperature drops a little, but the structure of the star hardly changes.

To conclude our discussion of the evolution of small stars we must note the following. Computations have shown that the star remains entirely convective during the main sequence stage if its mass is within the limits of $0.08M_\odot < M < 0.26M_\odot$ (at greater masses a radiative nucleus is formed). Since the central temperature is small here, the proton reactions stop with the formation of He^3 which can no longer turn into He^4. After ten milliard years 1 per cent. of the hydrogen is burnt, so that the ratio He^3/He^4 in these stars can be of about 3 per cent. Stars with $M < 0.08M_\odot$, having passed the stage of thermonuclear reactions (the temperature in their interiors does not reach eight million degrees), pass at once to the state of degenerate red dwarfs.

The evolution of massive stars is much more complex and interesting. We shall now study these stars. In massive stars, as we already know, there is always a convective nucleus containing 10–40 per cent. of the entire mass of the star. Hydrogen mixes rapidly here but also burns rapidly. However, it is not clear whether there is an exchange of matter between the convective nucleus and the surrounding envelope in which energy is transmitted by convection. It is more probable that there is no such exchange. In any case without this hypothesis we shall not obtain the stellar evolution scheme which agrees with optical data, so the question of mixing remains open.

Let us assume that there is no exchange of mass between the convective nucleus and the radiative region and let us consider what is the result. In the beginning hydrogen will burn uniformly over the whole convective nucleus.

Computations show that as the hydrogen content decreases in the centre of the star, the convective nucleus also diminishes by mass and by dimension. As the hydrogen content decreases from $X \approx 0.7$ to $X \approx 0.05$, the mass of the convective nucleus decreases by 2–3, and its dimension decreases even more due to an additional contraction of matter in the centre of the star. In this case the temperature in the centre increases by 10–15 per cent. and the central density increases by 15–20 per cent. The star remains on the main sequence as long as the temperature in its already small convective nucleus does not drop more than 1 per cent. What happens after the exhaustion of hydrogen in the centre of the star?

The energy flux from the stellar interior will not be compensated by thermonuclear reactions and the star starts to contract at least in its central part where the pressure had earlier started to decrease. Contraction of the star causes an increase in the central temperature. The temperature of the nucleus, which no longer contains hydrogen (it is now composed of almost pure helium), will also grow, as well as the temperature of the radiative part of the envelope surrounding the nucleus. Finally, the temperature grows so much that the hydrogen will burn in the central parts of this envelope. The star will now have the following structure: the central helium nucleus which is isothermic (i.e. at a constant temperature) is surrounded by a thin layer in which thermonuclear reactions occur due to the fact that here hydrogen is still conserved and the remaining envelope in which energy is transferred by radiation.

Can such a star exist? The reader probably remembers that similar structures are found in the inner parts of red giants although there is, besides the above described layers, also a very elongated outer envelope with convective energy transfer. Apparently such a convective envelope is indispensible.

With convection in the isothermic nucleus the density is increased, but in the surrounding layers where hydrogen burns the density should not increase. If this were to happen, then the energy release would grow sharply (it has large amounts of hydrogen and large temperatures and densities) and the transparency of the matter would decrease sharply. This cannot take place, however, because the amount of energy generated in the star is always regulated by the heat evacuation. The increase in energy emission with a simultaneous decrease in transparency leads to a violation of this principle. From this it follows that the increase in density of the central nucleus leads to a decrease in the gas density in the adjacent envelope. The isothermic part contracts and the layer with burning hydrogen remains in its place. Rarefications are not possible in a star; the density of the gas must always decrease with the emission from the centre. Thus, it appears that the formation of a layer with hydrogen burning surrounding the isothermic nucleus must inevitably be accompanied by expansion. In the outer layers of an expanded star the temperature drops, the transparency decreases, and these layers pass to the state of convective energy transfer. In short, after the beginning of hydrogen consumption in a layer the former main sequence star turns into a typical red giant with a complex inner structure.

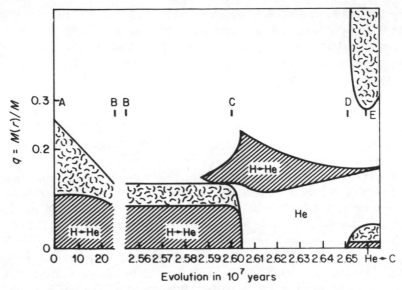

Figure 23 Change of the stellar structure in the process of evolution at the stage of hydrogen burning. Straight shading shows the regions of thermonuclear reactions; the figured shading shows the convective zones

These qualitative estimations could appear to be not very convincing to the reader. Simple numerical estimations cannot be done here, but numerous calculations of stellar evolution models have confirmed these conceptions.

Incidentally, in the process of passing to the stage of a red giant the luminosity of the star can for a certain time decrease since an important part of the energy produced goes towards the blowing-off of the stellar envelope and its 'boiling', i.e. the establishment of convection. In this case the surface temperature also sharply decreases. Afterwards, however, the luminosity starts to grow again.

Can such a layer described above convert the entire amount of hydrogen into helium and by this turn the star into an isothermic gas sphere? It appears that this is not possible. It is easy to consider that an isothermic gas sphere will be unstable since the inner gas pressure will not be sufficient to support the upper layers which, notwithstanding their weight, exert a high pressure on the interior layers due to the high temperature. In a gas sphere restrained by its own gravity the temperature must always increase from the surface towards the centre. For this reason the central isothermic nucleus cannot take over the whole star. Calculations show that if more than $0.21\mu^2$ of the stellar mass is concentrated into an isothermic helium nucleus, it starts to contract again independantly of whether or not there is hydrogen in the layer (μ is here the molecular weight of the matter outside the helium nucleus). The temperature in the centre of the isothermic nucleus will start to increase. When it reaches a hundred million degrees (or a little more) the triple alpha process starts—the formation of carbon from helium. Helium naturally burns first in the centre of

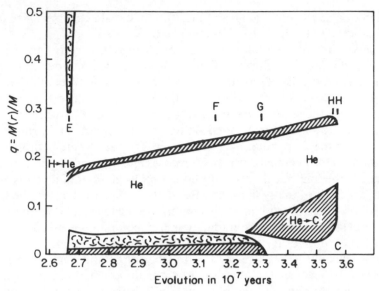

Figure 24 Stellar structure variation in the evolution process at the stage of helium burning

the star and our star again becomes similar to its initial model—the thermonuclear sources in the central convective nucleus. The convective envelope disappears and a hot, very bright star appears. All this is then repeated again, but for another stage. Helium burns in the nucleus and a star is formed with a layer of helium and a broad convective envelope. All that has been said above about a star with a hydrogen burning layer is also valid here. Only the concentration of density towards the centre of the star is even more distinct. The relative radius of the layer of helium is five times smaller than the relative radius of a hydrogen thermonuclear layer. The total radius is even greater and the star becomes a yellow or red supergiant.

At the end of helium burning in a layer a star is formed with the following distinctive chemical composition: in the central nucleus (22–16 per cent. of the mass) are carbon, oxygen, and neon; this nucleus is surrounded by a layer in which helium remains (about 3–5 per cent.); further out this is surrounded by an envelope basically composed of hydrogen (71–73 per cent. of the mass). Contraction then begins again, the temperature in the centre increases, and if the mass of the star is big enough other reactions can start—those dealt with in Chapter 4 (formula 19). First a star is formed with a central carbon source and then a star with a layer carbon source. This is no longer a simple 'repetition of what happened'. Neutron luminosity grows sharply and all processes are explosive. We shall see later what will happen.

The evolution of a star is well illustrated in Figures 23 and 24 drawn by R. Kippenhahn. They show the change in structure of a star with a mass of about $7M_\odot$ during its evolution. Along the vertical axis is given the part of mass which is occupied by a layer and along the horizontal axis the time. Since the evolution of a star is not regular in time, the time scale changes twice on the horizontal axis. Let us examine these figures in detail.

We have already seen that as hydrogen burns the convective nucleus diminishes; this is clearly shown in Figure 23, where the figured shading marks the convective zones. Subsequently the mass in the regions where the hydrogen burns also decreases slightly. The time on the main sequence is indicated by AC. At point B the scale of the time axis has been changed in order to extend the time scale to include red giants. At point C the hydrogen burning in the nucleus comes to a stop, but a layer source starts to burn where hydrogen turns into helium. In the beginning its relative mass is big, up to a few per cent. of the total mass of the star, but after that the layer rapidly grows thinner and contains only 1.5 per cent. of the stellar mass. Beneath this layer is located the helium nucleus. For stage CD nothing is burning in this nucleus and the temperature remains the same; however, it gradually increases with time. Figure 23 shows that the penetration of the convective surface zone into the interior nuclei of the star occurs only at the stage immediately before helium starts to burn in the nucleus (DE). According to other calculations the convective surface zone is formed earlier. At point D helium starts to burn in the centre and a convective nucleus appears again.

The further evolution of the star is shown in Figure 24 where the time scale is again a little compressed. Thermonuclear reactions take place in the star simultaneously in two areas: the helium reaction 'works' in the nucleus and hydrogen continues to burn in the layer. Note that, as could be expected, the hydrogen layer moves upwards through the mass of the star (i.e. it leaves behind an always greater part of the mass). In this case the part of the stellar mass in which hydrogen is burning remains more or less constant. Although the helium reaction in which three helium nuclei become one carbon nucleus gives a large energy release and there is enough helium in the nucleus, the lifetime of the high luminosity of a star with a helium source is about ten times shorter than that of a star with a hydrogen source. Figure 24 shows this clearly. Then helium is used up in the centre of the star and a layer helium source starts to burn (indicated by GH). A carbon nucleus appears in the star. The further stages are not marked in Figure 24, it is very difficult to calculate them.

Figures 23 and 24 clearly show the evolution of a star, but for a comparison with observations it is better to build evolution tracks, as for example that shown in Figure 25 of a star with a mass of $5M_\odot$, calculated by the American astrophysicist I. Iben. Figure 25 shows the variation of luminosity and surface temperature of a star in its evolution process, as well as diverse stages and the characteristic times which stars spend on each of these stages. The main sequence is shown by the track between points 1 and 2. Then follows a small phase of contraction (2–3) and the establishment of the layer source. The decrease in luminosity on the portion 5–6 is linked to an energy loss for the expansion of the convective envelope. The indications given below the figure allow the reader to examine all the phases.

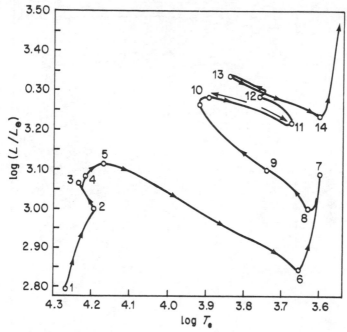

(1–2)—hydrogen burns in the convective nucleus, 6.44×10^7 years; (2–3)—general contraction of the star, $2.2. \times 10^6$ years; (3–4)—hydrogen starts to burn in the layer source, 1.4×10^5 years; (4–5)—hydrogen burns in the thick layer, 1.2×10^6 years; (6–7)—red giant phase, 5×10^5 years; (7–8)—beginning of helium burning in the nucleus, 6×10^6 years; (8–9)—disappearance of the convective envelope, 10^6 years; (9–10)—hydrogen burns in the nucleus, 9×10^6 years; (10–11)—second expansion of the convective envelope, 10^6 years; (11–12)—contraction of the nucleus during helium burning; (12–13–14)—layer helium source; (14–?)—neutrino losses, red supergiant.

Figure 25 Evolution track of a star $5M_\odot$ belonging to type I population stars. The particularities of each evolution phase are given below the graph, as well as the length of these phases in years

The evolution tracks for stars with different masses are given in Figure 26. The numbers on the turning points of the tracks have the same meaning as on Figure 25. Slowly evolving stars with small masses are indicated by dashed lines. Detailed calculations show that the evolution of stars with masses in the interval $0.1M_\odot < M < 3M_\odot$ have certain particularities after a certain time of evolution. First of all, if the stellar mass is inferior to $0.5M_\odot$, helium will not burn at all. This can be explained in the following way. We already know that as hydrogen burns in the central part of the star the gas density becomes greater and the temperature smaller. It appeared that in stars with masses below $0.5M_\odot$ the density is so big and the temperature so low (a few million degrees) that a degeneration of the electron gas occurs with contraction. Here the helium nucleus is formed of degenerate electron gas. Of course it continues to contract but in a degenerate gas the pressure is

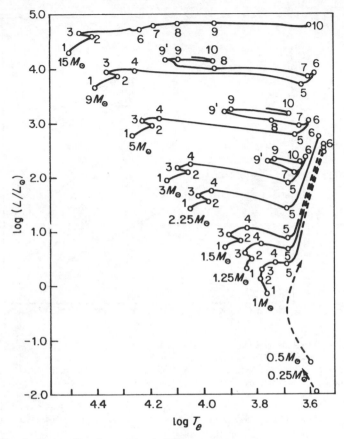

Figure 26 Evolution tracks for stars of different masses. The numbers indicate the same evolution phases as in Figure 25

determined only by the density and therefore the contraction is not accompanied by an increase in temperature. In such stars the temperature never reaches the hundred million degrees necessary for helium burning.

In stars with masses from $0.5M_\odot$ to $3M_\odot$ electron degeneracy also occurs on contraction of the central nucleus, but to a smaller degree. The temperature of the stellar centre increases with hydrogen burning but the increase is slower than in more massive stars. Eventually the temperature is large enough to make the helium burn. However, in order to compensate somehow for the slow temperature increase, helium here does not simply burn but 'flares' (the so-called helium flare). Let us explain this phenomenon. First we shall examine how helium starts to burn in massive stars which lack degenerate electron gas. As the gas is compressed the temperature grows and the gravitation pressure increases slowly. At a certain moment the temperature becomes high enough to allow helium burning. Energy starts to be released and the temperature grows a little more rapidly. This brings about an even more rapid increase of the gas pressure which seems to be greater than the gravitation pressure. The stellar nucleus rapidly starts to expand and the excess temperature decreases. In the end the stellar nucleus reaches, comparatively quickly, an equilibrium state at which the energy release is compensated by an outward flow.

In a nucleus with partly degenerate electron gas the situation is different. Let us assume here also that the temperature has reached a level at which helium starts to burn. A certain amount of energy has been released and the gas is strongly heated. However, the gas pressure has not changed very much as it is, at least partially, determined by the electron degeneracy and depends to some extent on the temperature. The nucleus continues to contract and the temperature continues to grow even more rapidly. This leads to an even more rapid helium burning and a greater temperature increase. Only at temperatures where the electron gas is no longer degenerate will the nucleus of the star start to expand. Meanwhile, the energy emission due to the helium reaction increases strongly. In fact, an outburst takes place which is called a helium flare. This is only theoretical; it is still not absolutely certain how this occurs in reality. It should be recalled that stars with small masses evolve slowly and therefore what has been said above concerns type II population stars which have left the main sequence. In type I population stars only those with masses greater than 1.2–$1.5 M_\odot$ manage to leave the main sequence.

Stellar evolution computations determine the time dependence of different parameters characterizing the star. We cannot give here the tables of these detailed data, but certain tables have been included which were obtained by different authors on the basis of numerous calculations. Table 6 shows the characteristic lifetimes (in years) on the main sequence of three groups of stars with different masses. The first group (second column) are young type I population stars which on leaving the main sequence have many heavy elements ($Z = 0.02$) and a high abundance of helium ($Y = 0.3$). The second group also belongs to type I population stars but these are older stars of intermediate systems. There are less heavy elements ($Z = 0.01$) and less helium ($Y = 0.1$). Finally, the third group is composed of type II population stars of spherical subsystems. In calculations the amount of heavy elements has been assumed to be very small, about 10^{-3}–10^{-4}. The dashes in the table indicate that such stars have not yet been considered. Finally, Table 7 gives the characteristic lifetimes of stars in the red giant stage. For type II population stars only those with small masses are considered, as they are the only ones which remain today. It should be pointed out that the calculations of different authors often differ.

At the present time diverse institutions continue intense work on the calculation of stellar evolutionary sequences. The results of these calculations are in good qualitative and often quantitative agreement, although they still differ in details. Apparently it will soon be possible to compose detailed tables which will permit determination of the values of the basic stellar parameters M, L, and R at each given moment depending on the initial conditions of their formation.

Table 6

Stellar masses in units of solar mass	Stars of normal composition (plane subsystems)	Stars of intermediate subsystems	Stars of spherical subsystems
64	2.5×10^6	—	4×10^6
32	4×10^6	—	6.6×10^6
16	8×10^6	1.5×10^7	1.0×10^7
10	2×10^7	—	2.5×10^7
6	7×10^7	1.2×10^8	1.1×10^8
3	2×10^8	—	—
1.5	1.5×10^9	—	—
1.0	7×10^9	2×10^{10}	—

Table 7

Stellar masses in units of solar mass	Stars of normal composition (type I population)	Stars with poor amount of heavy II population)
64	5×10^3	—
32	1.5×10^4	—
15	2×10^6	—
10	4×10^6	—
5	2×10^7	—
3	7×10^{10}	—
1.5	4×10^8	10^{10}
1.0	4×10^9	2×10^{10}

Let us go back to Figures 25 and 26 and compare the theoretical calculations with observational data. We immediately see a good explanation of the location of the giant branch in globular clusters. They are composed of old stars and therefore the red giant stage has already been reached by stars of small mass $0.5M_\odot < M < 1M_\odot$). Figure 26 shows that their calculated evolution follows the vertical line on the spectrum–luminosity diagram in perfect agreement with observations. These stars are at the stage of a layer hydrogen source.

We know the position of the zero age stars. Now let us mark on the spectrum–luminosity diagram the position of all stars within any determined time interval and join the points by a line. Thus we obtain the position of stars of the same age. This has been done in Figure 27 for stars of the upper part of the main sequence and for relatively small time spans (hundreds of millions of years). In Figure 28 this has been done for the entire main sequence and for larger time spans (several milliards of years).

The properties of these sequences with different ages are easy to understand. Because of the increase in luminosity and the decrease in surface temperature, occurring more rapidly in massive stars, the sequence of stars of the same age goes from the main sequence of zero age upwards and towards the right-hand side. The passage to the red giants interrupts these sequences at their ends and the older the curve the lower this interruption—less massive stars have always managed to become red giants.

Now let us compare Figure 27 with Figure 3. The rules appear clearly. The young cluster NGC 2362 just managed to 'emerge' on the top of the main sequence of zero age. The cluster in the Perseus constellation, which is older, 'merges' even more and now appears on a big branch of bright red giants. The clusters M41 and M11 are even older and so the point of departure from the main sequence is lower. The red giants are situated below here—this is linked to the fact that massive stars went through the red giant stage and the observed red giants are already stars of smaller masses. There is a very good agreement between theory and observations here!

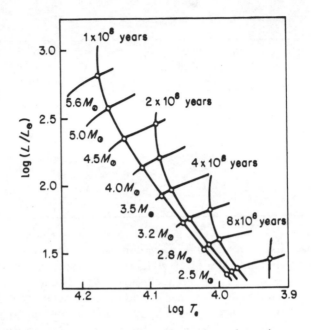

Figure 27 Displacement of stars of different mass from the main sequence with time

In the very old type II population clusters M92 and M3, as well as in the old type I population cluster M67, the evolution has at first sight a slightly different character: here the giant branch is not interrupted and leaves the main sequence, or the subdwarf sequence, at a right angle. In fact, the evolution is the same, although slower, and concerns stars with smaller masses. Because of this slow evolution the stars gradually pass to red giants without passing through the intermediate region, as is the case for massive stars.

In Figure 28 sequences of stars of two old clusters are shown: one being the M67 cluster, already known to us, and the other the very old NGC 188. Comparing the theoretical positions of stars of the same age with the observed sequence we can also determine the age of the corresponding cluster. For example, the age of the cluster in the constellation Perseus is probably lower than ten million years, the age of the Pleiades is about a hundred million years, the age of the Hyades can reach one milliard years, the age of M67 is about eight milliard years, and the age of the cluster NGC 188 appears to be of the order of cosmological time—about 14–15 milliard years.

Thorough spectral studies have allowed us to obtain the so-called 'fine structure' of the main sequence and also to determine the age of different stars. Figure 29 shows an example of such a structure (obtained by B. Strömgren). The farther the star is from the sharp line of zero age (the lower part) the older it is.

Thus for the general evolution of stars the passage from the main sequence to red giants is in very good agreement with the observational data. Many details are also in agreement. For example, the fact that on Figure 3 the type II

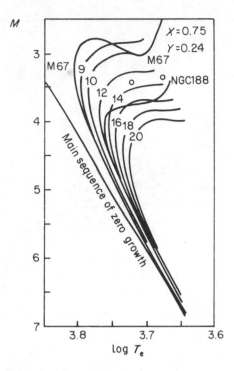

Figure 28 Displacement of stellar sequences from the main sequence with time. The lighter lines indicate the calculated positions of the sequences at different moments (the time spans are given in milliards of years) and the heavy lines show two observed sequences

population red giants are situated on the spectrum–luminosity diagram above the type I population red giants can be explained by the fact that in stars with a small amount of heavy elements the opacity is small and consequently the luminosity is high. Other, finer properties are linked, for example, to the fact that, in stars with small mass, on the formation of a nucleus without hydrogen this nucleus is made up of degenerate electron gas; this limited the increase in temperature and does not lead to a helium reaction flare. It also leads to a continuous transition from the main sequence to the red giants.

This scheme explains the evolution of ordinary single stars. However, we know that, first, there are sufficient stars with such properties, as for example variable stars. Second, a great number of stars are double and, in general, form a multiple system. The theory of evolution should also consider these stars.

Comparing Figures 25 and 26 with 12 we can see that pulsating variable stars are located on the spectrum–luminosity diagram in an intermediate region between the main sequence and the red giants. For this reason one can expect that the capacity of a star to pulsate is linked to a definite stage of its evolution when, as a result of the expansion of the outer stellar layers, there is a situation in which helium and hydrogen ionization zones oscillate with a 'negative dissipation'. In fact, calculations confirm this assumption. The envelope

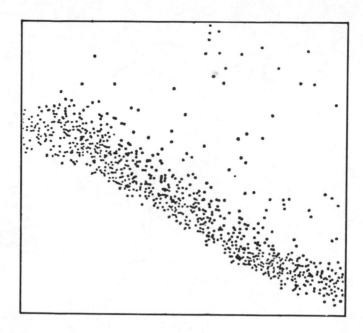

Figure 29 'Fine structure' of the main sequence. Here special colour indexes are chosen to characterize the presence in stellar spectra of a line of metals (horizontal axis) and of hydrogen (vertical axis)

expanded during the formation of a helium nucleus is unstable with respect to the excitation of pulsations if the helium abundance is not less than $Y \approx 0.22$.

Detailed calculations allow us to follow the evolution of a star entering the phase of pulsation excitation and, after a certain period (usually 10^4–10^5 years), leaving this phase when the pulsations stop. The luminosity of the star at this stage of evolution determines the dimension of its envelope and consequently the mean density and the pulsation period. From this the period–luminosity relationship is deduced which is described by the graph in Figure 13 obtained with observational data.

Now let us study the evolution of binary stellar systems. First of all we must say that until now the cause of the formation of binary systems remains unknown. The question is how, from an initial fragmented cloud, pairs of stars can be formed which almost touch upon each other and rotate around a common centre of gravity. Many hypotheses exist but none of them are very satisfactory.

If the distance between two stars of one system is very large, e.g. a thousand times greater than their radii, then each star evolves by itself and there will be no particularities in the development of binary stars. If, in this case, the masses of the components of the pair are very different, the more massive star evolves more rapidly and, in such a pair, two stars at different stages of evolution can be linked, e.g. a star of the main sequence and a white dwarf. However, if the stars form a close pair the evolutionary course is altogether different. In this case one star can outweight the mass of the other star. Stellar evolution in close binary systems has been given much consideration and we shall describe certain results which have been obtained.

First we shall introduce the idea of the Roche lobe, which is very important for the entire evolution theory of binary stars. For this we shall study the motion of a particle, appearing for some reason in the gravitation field of both stars. If the particle is close to the surface of one star it will be attracted to this star, but if the particle is somewhere between the stars, or even simply nearby, the attraction from each star will partly or entirely compensate each other and the motion of the particle will depend on the magnitude of the centrifugal force acting upon the particle, since it rotates together with the binary system of the stars. This means that stars in such a binary system can retain particles only if these are situated inside a certain closed region where the attraction of one star prevails over the attraction of the other star and the centrifugal force. This region is called the Roche lobe. The Roche lobes of both stars touch each other on a line linking the centre of both stars at a point which is called the first critical Lagrangian point L_1.

Coming back to the evolution theory of binary stars we can say that, evidently, all depends on how deeply 'embedded' the stars are inside the Roche lobes. Will they be able, during the whole time of their evolution, to extend so much as to fill the Roche lobes, even if only for a certain time? In fact as soon as the star fills in its Roche lobe, its outer layers are no longer gravitationally related to the rest of the star. The star loses its mass and this must appear in the course of the whole evolution of the star. There is not only mass loss when the star fills its Roche lobe. Part of this mass, namely that which leaves the lobe close to the Lagrangian critical point, is captured by the second star, falls on its surface, and so changes the evolution pattern of the second star. This phenomenon is called accretion. These properties—mass loss from the first star and accretion of part of it by the second star—determine the whole evolutionary pattern of binary systems.

It is evident that the filling up of the Roche lobe can take place only if it is not too big, i.e. if the distance between the stars is comparable to the dimension of these stars. For this reason only close binary systems have distinct particularities of evolution.

It is well known that there exist close binary systems in which the distance between the components of the pair are five to twenty times greater than the solar radius. If the stars of such pairs have masses greater than that of the Sun and subsequently also have greater radii they can fill up their Roche lobes during their stay on the main sequence. This is called a type A evolution. On the other hand, if the distance between the stars of one pair is greater than the radius of the Sun, e.g. thirty to two hundred times (the radius of the Sun being $R_\odot = 7 \times 10^{10}$ cm), then the star can only fill up its Roche lobe at the red giant stage when there is a helium nucleus and a layer hydrogen source of thermonuclear energy. In this case it is a type B evolution. One more expansion of the star occurs at the stage of hydrogen burning in the layer source. This is a type C evolution.

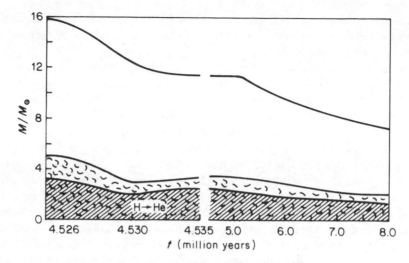

Figure 30 Change in structure of a star belonging to a close binary system. Mass loss in the evolution process at the stage of hydrogen burning in a convective nucleus (Graph calculated by A. Tutukov and L. Yungels)

At each stage of expansion the star loses mass which leads to an increase in the distance between the stars. Therefore one pair of stars can at the beginning evolve as a close pair (i.e. at stage A) and then as a larger pair (stage B).

Let us first consider the evolution of a close binary system where the main component fills up its Roche lobe already at the main sequence stage. Calculations show that as soon as the stellar surface approaches its Roche lobe in the expansion process there is an outflow of matter. The first stage of outflow lasts about 10^4 years and during this time the star loses 15–20% of its initial mass. Then its surface recedes again from the Roche lobe and the outflow stops. The star somehow accommodates to a new existence of a so-called subgiant where the mass of the star is smaller than that corresponding to its given luminosity. We should note that in the process of outflow the luminosity of the star changes very little since it is determined by the magnitude of the energy sources and the radiative transfer in the deep layers of the star where the equilibrium is not very much affected by the outflow of matter from the surface. Afterwards the star with a lesser mass, being on the main sequence, expands again, fills up its Roche lobe again, and starts again to lose mass—but now much more slowly. In two to three million years it loses about 25–35 per cent. of the initial mass. At stage A a star of a close pair can lose up to half of its initial mass.

Figure 30 shows a graph representing the change in structure of a star—a component of a close binary system during the process of mass loss. The vertical scale indicates the stellar mass expressed in solar masses at each moment in time (with an initial mass of $16M_\odot$) while the horizontal scale indicates the time. The time scale changes in order to describe the slower mass loss during the second stage. We can see in Figure 30 that the convective

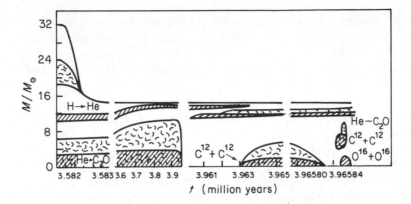

Figure 31 Change in structure of a massive star during its evolution process. Fundamental mass loss at the stage of hydrogen burning in the layer source and helium burning in the nucleus. In the later evolution stages, carbon and oxygen burning are also shown (Graph by A. Tutukov and and L. Yungels)

nucleus and the region of hydrogen burning change little with the outflow of matter, although during the intermediate period when there is an absence of outflow a certain change in structure occurs and there is an accommodation to the smaller mass.

In case B, when the star fills up its Roche lobe, the pattern will be somewhat different at the stage of the layer source, although in this case there are also two stages of mass loss. In the beginning a rapid outflow of mass starts as soon as the stellar envelope fills up its Roche lobe and the star loses 50–70 per cent. of its mass within 10^3 years. During the second, slower phase, the star loses about one solar mass within about 10^5 years, since all mass outflow during stage B, where two-thirds or even more of its mass leave the star, is ten times more rapid than in case A.

Figure 31 represents the construction of a massive star (with an initial mass of $32M_\odot$) during its evolution process if it belongs to a binary system. Here only late evolution stages are shown, helium burning in the nucleus, then a helium layer source, and carbon and even oxygen burning in the nucleus of the star. In the left-hand side of the figure the large mass loss during stage B is shown which is followed by a relatively small but rapid mass loss (within 10^3 years) at stage C—a loss of about $0.2M_\odot$.

From this we can obtain a formula which will determine the mass of the remainder of the star after all outflows of mass M_f relative to the initial mass M_1 have been considered.

$$(59) \qquad \frac{M_f}{M_\odot} \approx 0.1 \left(\frac{M_1}{M_\odot}\right)^{1.4}$$

The greater the initial mass the greater the amount of mass loss.

The outflow of matter in close binary systems leads to the fact that the component stars are often particular objects. Let us study some of them. For example, astronomers have known for a long time the existence of so-called Wolf–Rayet stars which are characterized by a rather rapid mass loss. Observations show that these stars lose 10^{-6}–$10^{-7} M_\odot$ in a year. It also appears that these stars enter close binary systems and that the mass of Wolf–Rayet stars is about three to five times smaller than the mass of the second component forming the pair.

However, the most interesting fact lies elsewhere. It appears that all Wolf–Rayet stars are divided into two groups: nitrogen and carbon stars. In nitrogen stars the amount of nitrogen is fifty to a hundred times greater relative to the amount in ordinary stars and there is comparatively little carbon. In carbon stars, on the contrary, there is four hundred to seven hundred times more carbon than its relative amount in ordinary stars. The amount of hydrogen is relatively small in Wolf–Rayet stars, particularly in nitrogen stars.

Can these particularities be explained by stellar evolution theory? It appears that at least qualitatively this can be done quite easily. Let us consider the stage of evolution of a star when after the helium consumption a carbon nucleus is formed in its centre. There is little helium (it was used in thermonuclear reactions), there is absolutely no hydrogen, and there is a relatively small amount of nitrogen. Above this nucleus is a layer of matter with a lot of helium but almost no hydrogen; it was also used in the hydrogen burning, mostly during the carbon–nitrogen cycle (17).

As we have already seen, the result of these reactions is the following relative abundance of nitrogen and carbon: about 95 per cent. of the nitrogen isotope N^{14}, about 4 per cent. of the carbon isotope C^{12}, and only about 1 per cent. of the carbon isotope C^{13}. Consequently, in the layer of material surrounding the carbon nucleus nitrogen and helium prevail. In even higher layers some hydrogen remains and the ratio of carbon to nitrogen is more or less normal.

We now assume that such a star with a layered composition starts to lose mass as it enters a close binary system and fills up its Roche lobe. In the beginning only the upper layers will flow out and we see a star of normal composition. Then the second layer starts to flow—a region rich in nitrogen which remained below the layer of burning hydrogen. This layer also flows out entirely. As a result layers on the surface appear to be enriched by carbon on account of the burning helium. Wich such a gradual uncovering of deeper and deeper layers of the star which differ by their chemical composition we can explain the existence of nitrogen and carbon Wolf–Rayet stars. In the process of mass loss the distance between the stars increases and in fact the carbon Wolf–Rayet stars are located farther from their companion than a nitrogen star. Of course the Wolf–Rayet stage can be passed only by stars with a great initial mass, of about 5–10 $M_\odot$, which is understandable since most of their mass has been lost.

We must note that single stars also lose mass if in the process of their evolution they increase their dimension so much that the velocity of the matter flying off their surface is of the order of the thermal velocity of molecules. The outflow of matter can also be favoured for the pressure of radiation if the luminosity of the star becomes very strong in the process of evolution. A so-called Eddington limit of luminosity exists:

$$(60) \qquad L < 10^{38} \left(\frac{M}{M_\odot} \right) \qquad \text{erg/s.}$$

It is obtained in the following way. We compare the force which acts upon a particle of matter absorbing the radiation from a star with the action on the same particle of gravitational attraction forces and we find the luminosity at which these forces are equalized. Let us explain this in more detail. The light radiated by the star transports an impulse—a certain amount of motion—and when a particle of matter absorbs this radiation it also takes away the impulse and consequently receives a shock in the direction of the light emission, i.e. from the star. The force of the light decreases in inverse proportion to the square of the distance and thus changes the impulse absorbed by the matter. The gravitation force also decreases in inverse proportion to the square of the distance and therefore if the repulsion force is greater, due to the absorption of radiation, than the gravitational attraction this condition is satisfied at all distances from the star. Formula (60) determines this luminosity where the repulsive force on the absorption of radiation is greater than the force of attraction. It is clear that during the evolution of a star the luminosity reaches the limit (60) and the outer layers will be torn off by radiation pressure and fly off into infinity. It is probable that in this way, for example, single Wolf–Rayet stars are generated which are also sometimes met.

Astronomers often deal with mass losses by stars. Sometimes the rejected mass can be observed, e.g. planetary nebulae. Probably they were formed by red giants in which the whole envelope gradually left the star. The slow expansion of the envelope is also observed in the form of a spherical layer surrounding the remainder of the star, the so-called nucleus of the planetary nebula which from its properties slightly resembles Wolf–Rayet stars. The whole process of tearing off and recession of the envelope, although smooth, is rather rapid and therefore a few planetary nebulae are observed—in our Galaxy there are about one hundred.

Other stars lose mass in a not very quiet manner. Such loss is observed in the form of outbursts or stellar flares. These flares can be of very different magnitudes. We have already described these processes and how they are observed in Chapter 9.

Unfortunately it has not yet been possible to reconcile the phenomena occurring in stellar flares with the evolution theory. The masses rejected in flares of nova-like and nova stars are small and therefore such flares apparently are not related to any important reconstruction of stars. Probably they occur on a more advanced evolution stage and apparently an important part is played by the recently discovered duplicity of such stars. This duplicity acts upon the stability of such stars in a way which is still not very clear. The reader should not be astonished that there is so little to say about these stars. When we studied the evolution of ordinary stars and even binary stars with mass loss, we considered that a star is at every instant in equilibrium state and we assumed that only slow changes occur with time in the state of equilibrium. Even pulsating stars oscillate around their equilibrium state. Flaring stars are not in equilibrium, all processes are rapid, and it is much more complicated to calculate them. For this reason we have very little data but we hope that in the near future more information will be found on these interesting objects. We shall come back later to flares of supernovae stars.

Studying the evolution of stars, single or double, we have seen many times that during their evolution a gradual formation of heavier elements takes place in all stars—in the beginning in the central parts and then in higher layers. The question arises: up to which elements can this process of formation reach? The answer is related to the estimation of the maximum temperature that can be reached in stars of different mass in the process of the whole evolution.

Although it is difficult to effect reliable calculations, according to estimated data the end of the thermonuclear evolution of stars with masses greater than $4M_\odot$ occurs when a nucleus is formed in the centre of the star which contains carbon, oxygen, neon, and metals (such as magnesium and iron). In this 'metallic' nucleus 20–25 per cent. of the mass of the whole star is concentrated. The nucleus is surrounded by a layer in which much unburnt helium remains (about 5 per cent. of the mass). If the star conserves the greatest part of its mass, then its outer layers are composed, as before, of matter with a large amount of hydrogen.

A star with a nucleus composed of heavy elements has no longer a sufficient amount of thermonuclear fuel for corresponding reactions in the region where the temperature is sufficiently high. What happens then? Evidently a star deprived of thermonuclear sources starts to contract and its central parts pass to the state of electron degeneracy (if it has not already taken place at an earlier stage). What happens next depends on the mass of the star and on how much mass is lost. We assume that the mass of the star was below the Chandrasekhar mass limit of white dwarfs, but smaller than the mass limit of neutron stars. Therefore, in the contraction process the stellar gas is neutronized and a neutron star appears. It can be a pulsar if it rotates rapidly and has a strong magnetic field. However, we are not sure that all neutron stars become, at least temporarily, pulsars. In fact it is known that among stars there exist many binary systems and these are few among pulsars. Apparently, only the stars which do not accrete mass on their surface from the second components of the pair are converted into pulsars. Finally, if the mass of the remainder of a star is sufficiently large it will in the process of contraction recede into its 'black hole'.

As we already noted, the process of rejection of envelopes by stars, in particular after exhaustion of thermonuclear sources in the central parts of the star, has not yet been studied in enough detail. Some calculations were done, however. Here it is very important to consider the role of neutrinos. In fact, the contracting nucleus of a star becomes very dense and hot and therefore nuclear processes with neutrino emission can take place. If they leave the star freely, then the energy they transport decreases the gas pressure and facilitates contraction. However, if the neutrino is absorbed in the upper layers of the contracting star, then the energy is transmitted to these layers which leads to the rejection of a certain enevelope. This can happen with stars having masses smaller than $2–4M_\odot$.

In the case where the nucleus becomes so dense that even neutrinos cannot leave it, once the density of neutron gas is reached the contraction of the nucleus will be suspended. The higher layers continue falling, meet the nucleus, and a strong shock wave appears. In the collision with the dense nucleus the gas is heated. Because of this, thermonuclear reactions can again appear in the falling matter of the outer stellar layers. The temperature here increases so much (up to 10^9 degrees) that not only helium and hydrogen are rapidly and totally burnt (if these elements are still conserved) but oxygen also starts to burn. A very large amount of energy is emitted and the most external envelope of the star is rejected. According to calculations, this phenomenon can happen in stars with masses smaller than $15–20M_\odot$.

This process of envelope rejection due to oxygen burning in the outer layer of such a star is, generally speaking, linked to the phenomenon of flares in supernovae. Of course, as the same calculations show, the masses and the energy of the rejected envelopes are smaller than the magnitudes observed in real supernovae.

If the mass of the star is very large ($M > 30M_\odot$) then nothing can stop its rapid contraction after the combustion in the nucleus of all thermonuclear energy sources. Even the remaining unconsumed oxygen, helium, and hydrogen in the outermost layers of the star do not manage to explode. Everything recedes into the 'black hole'.

These conclusions were obtained through calculations which did not take into account the role of rotation: its calculation complicates the problem. On the other hand, as we already know, during contraction the rotation increases strongly which can lead to an additional rejection of the envelope due to the action of the centrifugal force. This problem is in general very complex and we will probably have to wait some time for a more or less reliable solution. There is also little observational data on supernova flares or on this stage of stellar evolution.

The very last stage of stellar evolution—white dwarfs, neutron stars, and 'black holes'—were studied in Chapters 5, 6, and 7 of this book. The matter rejected by stars forms new stars but what has fallen into the 'black holes' is lost for further evolution.

The study of stellar evolution has always been one of the most important problems in astronomy. Many fantastic assumptions have been made. The first scientific hypothesis appeared when the American astrophysicist G. Russel, having built the spectrum–luminosity diagram, assumed that stars evolve on it along the main sequence from above to below. The modern pattern of stellar evolution is quite different from the first models. It has been realized from the work of many astrophysicists and is based on the theory of stellar structure described in this book. Many astrophysicists have published a modern pattern of stellar evolution, but since this is a collective work of scientists we shall not give particular names here.

Index

154